Re-Imagining Development Communication in Africa

Re-Imagining Development Communication in Africa

Edited by Chuka Onwumechili
and Ikechukwu Ndolo

LEXINGTON BOOKS
Lanham • Boulder • New York • Toronto • Plymouth, UK

Published by Lexington Books
A wholly owned subsidiary of The Rowman & Littlefield Publishing Group, Inc.
4501 Forbes Boulevard, Suite 200, Lanham, Maryland 20706
www.rowman.com

10 Thornbury Road, Plymouth PL6 7PP, United Kingdom

British Library Cataloguing in Publication Information Available

Library of Congress Cataloging-in-Publication Data
Re-imagining development communication in Africa / edited by Chuka Onwumechili and Ikechukwu Ndolo.
p. cm.
Includes bibliographical references and index.
ISBN 978-0-7391-7614-6 (alk. paper) — ISBN 978-0-7391-7615-3 (electronic) 1. Communication in community development—Africa. 2. Communication and technology—Africa. 3. Social change—Africa. 4. Mass media in community development—Africa. 5. Technological innovations—Social aspects—Africa. I. Onwumechili, Chuka. II. Ndolo, Ikechukwu S.
HN780.Z9C63 2012
302.2096--dc23

2012032535

™ The paper used in this publication meets the minimum requirements of American National Standard for Information Sciences—Permanence of Paper for Printed Library Materials, ANSI/NISO Z39.48-1992.

Printed in the United States of America

Contents

Preface

Scholarship in development and social change has been contentious for decades, particularly after the seminal work by Daniel Lerner when he proclaimed the passing of traditional society and noted that development would arrive through modernization. The criticism and the debate that followed Lerner's work produced volumes of work with a significant number of them focused on the situation in the newly independent African states in the 1960s. While much of that debate has finally slowed down, there is no reason to believe that it has died. The United Nations initiation of Millennium Development Goals and the increasing presence and participation of non-governmental organizations (NGOs) provide space for continued study and understanding of development and social change.

This text is a new look at development and social change communication. The title *Re-Imagining* is appropriate and deliberately chosen to highlight the fact that we need to take a fresh and new look at development communication in the Africa region. However, taking such a fresh look does not entail abandoning the past. This is important as we strongly believe that we cannot achieve an effective future without clearly understanding the past and learning from it in order to proceed.

In addition, we have a mixture of scholars and practitioners who contribute to this book. The scholars are both experienced and new. The experienced scholars were involved in the great debates of the 1970s and 1980s and the new ones are gearing up for new debates. There are some practitioners who have contributed chapters to help us bridge the gap between the academy and practice. Ernest Waititu, for instance, is a program manager for health and digital media at Internews in Nairobi, Kenya. His closeness to development practice in Kenya is a significant contribution to the country case in chapter 12. Genet Lakew is a professional who works for a non-governmental organization focusing on East Africa region. This mixture of chapter contributors provides impetus for fresh and new ideas.

Importantly, the book puts all the above into consideration in its organization and its delivery. The book is organized into three sections or parts, the first focusing on the past and the history of development communication scholarship; the second analyzes theoretical issues; and finally a third section that looks at country cases. We briefly discuss each of the sections below.

The first section begins with Daniel Awodiya's historical recollection of development communication debates. He begins his review by describing traditional perspectives led by Daniel Lerner's much-cited *Passing of Tradional Society* published in 1958. He also notes litany of criticisms that followed the works of Lerner, Schramm, and Pye among others. Chief among those critics was Paulo Freire. The chapter also discusses other paradigms including social marketing, cultural imperative, social mobilization, indigenous knowledge, and non-governmental organizations. For each of those perspectives, Awodiya provides both a description and a critique. In addition, he elaborates on one of the newest paradigms which is the information and communication technologies (ICTs) and their use in seeking development. Awodiya reviews several literature on the ICT paradigm in development and provides a critique that points to both strengths and weaknesses of the ICT perspective. Ultimately, he claims that the field has come full circle in paradigms about social change.

Onwumechili and Ndolo argue, in chapter 2, that while the development debate had pitted an economist-focus against a humanist-focus, the issue of development communication simultaneously pitted an indigenous media preference against modern mass media, particularly in Africa. The authors identify key changes in Africa since the 1960s when most African countries attained political independence. They note that most of those countries sought development in hopes of catching up with the West. They also note the failures that followed and how media role was conceptualized. In the later part of chapter 2, they argue that recent changes call for a new look at development communication. For instance, they note that urban population forms 34 percent of the national population in several African states, an increase from just about 11 percent in the 1950s, and this urban share is expected to outstrip rural population share in less than two decades. The authors note wide adoption of democratic system in Africa and emergence of new communication technologies. The authors argue that these changes portend a new urban audience and need to begin to focus on the use of emerging new media.

Duru's chapter looks at the role of African feminism in national development discourse. She points to a misunderstanding of African women in Western literature and notes that even in African traditional culture that several systems exist to empower women. She provides several examples of women resistance and activism that have helped empower women in Africa and further development. She also notes that Western literature often views African women as oppressed but she argues that in the contrary, the role of African women is complementary in several areas including economics and trade, and family.

Mogekwu argues in chapter 4 that in spite of the emergence of new digital communication systems, that the focus of development communication in Africa should be indigenous knowledge systems. He structures his argument by revisiting the history of development communication concepts from the early years of independence and pointing out the place of indigenous knowledge. Mogekwu carefully navigates the belief that such systems are static by providing examples of the dynamism of indigenous knowledge using examples from conflict resolution systems in African communities and healthcare practices, and restorative justice system such as practiced in the South Africa Truth Commission post-Apartheid. He suggests that such systems may be communicated using both what Ugboajah called *Oramedia* as well as new media such as the Internet and by instituting indigenous knowledge in educational curriculum.

The second section includes four chapters that revolve around key theories in the field. The first one, chapter 5, is written by Onwumechili and M'Bayo who provide a critique of the much cited Denis McQuail's theory of development communication system. Chapter 6 applies the theory of communicative action to development communication messages and chapter 7 uses McQuail's development communication to explain development communication practice in Ethiopia. The final chapter in the section focuses on a variety of theories including the stages of change theory and diffusion of innovation theory.

The critique of McQuail's theory in chapter 5 produces a new model, the Chumba development media theory (CDMT). The authors identify studies in Africa that challenge the validity of the development media theory as espoused by McQuail. Furthermore, they challenge the theory's assumptions that only the government drives national development and the fact that the theory largely ignores the importance of culture. In addition, they note that social changes have made McQuail's theory obsolete. It is based on those criticisms that the authors advanced the CDMT. The CDMT recognizes several sources of development communication and several modes for delivering such communication. The authors advance the CDMT using twelve theorems. The last section of their chapter is devoted to differentiating the CDMT from other theories of the press.

The next chapter by Pratt and Jacobson introduces the use of theory of communicative action in effective development communication. Pratt and Jacobson argue that African countries must produce effective communication for public awareness in order to respond to rapid population growth. Though the authors acknowledge the Western origin of the theory, they point out the theory's consistency with what they describe as "African rhetorical use of a conversational mode." They also note that what is prevalent is strategic action but that such action often exacerbates power inequalities. On the other hand, communicative action promotes conversations that denote interaction, negotiated compromises, equality, and participation.

Genet Lakew's chapter reminds us that the use of development communication theory advanced by Denis McQuail is ongoing in places such as Ethiopia which is the focus of Lakew's chapter. Lakew, a practitioner, provides detailed cases of the theory's applications to development communication in Ethiopia with examples from a health and then an education campaign. She notes that the Ethiopian government's draconian attitude towards media reports is focused on reports in two restricted issues of governance and human rights. Beyond those two areas, communication space is provided even to non-government sources. It is within this understanding that the Ethiopian government focuses attention on development communication and allows NGO participation. Lakew provides case examples of such development communication. One is on HIV-AIDS and the other is a children's education animation program titled Abeba and Abebe.

In chapter 8, Ndolo and Ozoh discuss the importance of development communication to reduce infant mortality rates in Nigeria. They argue that such communication must be effective at the individual, household, and community levels. They noted that achieving such effective communication at those levels requires the use of stages of change theory that begins from precontemplation to maintenance. They also advocate for Everett Rogers' diffusion theory and the social network and social support theories. These theories are connected with the fundamental and participatory process involving assessment and analysis before communication. Here they point out that the communication must be *sticky*, i.e., communication that is memorable because of its ability to mix several elements such as music, language, and color, among others.

The final section of the book focuses on country cases in Nigeria, Cameroon, Kenya, and Rwanda. While some of the chapters in the previous sections also focus on country cases, they were not included in this third and final section because they had theoretical frameworks that significantly guide the cases that they report. In the third section, the chapters emphasize cases without theoretical framework provided as a strong focus.

Ojo and Servaes in chapter 9 focus attention on an innovative approach to health development communication. They suggest the use of *Nollywood* movies to promote effective approaches to communicating health issues such as sickle cell disease. *Nollywood* movies have become important in recent years with the industry rising to number three in the world and shifting local focus from foreign movies to local ones. The authors provide two insightful examples to demonstrate how to effectively use movies for social awareness and change. In one case, they analyze the home video *Mortal Inheritance* which focuses around a female sickle cell patient who through her activities educates viewers about the disease. The sickle cell was simply a storyline of the movie but the main focus was entertainment. Ojo and Servaes criticize the movie claiming that the movie's opulence distracts from its educational significance. In the other case, *The Family Legacy*, unlike *Mortal Inheritance*, was clearly developed to create health awareness of sickle cell. Thus, the movie provides educational scenes surrounding various aspects of the disease, its myths, and misconceptions. Ojo and Servaes argue that the movie has a better chance of producing social change because of its strict focus on the disease.

Chapter 10 focuses on Cameroon, and specifically on Kumba. Kumba is a major area in the Anglophone region of Cameroon. Importantly, for us, it provides an interesting look at community conflict and the outside-in development project that has been a source of enduring conflict in the field of development communication. The Cameroon case points to several issues. A major one is that community that is not part of finding a solution to its internal development problems may not commit to the implementation of a solution. Thus, while it may be illogical to an outsider that communities may reject a bridge that would have brought development to them, for the communities the fact that they themselves were not integral to finding a solution to their problems is enough not to commit to a bridge that others may consider vital. Second, the chapter provides us with examples of using theater for communicating for social change. Theater is a good medium for such communication but it must work well with other variables in order to ensure development implementation. In this case, it remains arguable that it was optimally effective because of the communities feeling of not fully participating in planning for its own development.

Rwanda's *Gacaca* courts have been widely covered in the literature since the 1994 Rwanda genocide. In this chapter, Mutua and Tibeso provide a report on remarkable culture shift in Rwanda that was an outcome of the genocide. The culture change was deep and extensive—affecting traditional court system and the wider political system in the country. Rwanda was largely patriarchal society pre-1994 genocide. The genocide led to several deaths and abuses against children, women, and men. In the end, the patriarchal society changed. Women found spaces to participate significantly in the *Gacaca* courts and lead its administrative structure and women dominated the country's legislative house. The fact that women were in the majority in Rwanda's legislative house is in itself remarkable as such a distribution in a nation's legislative house exists only in Rwanda. In other parts of the world, men dominate legislative houses.

Gathigi and Waititu contribute an interesting chapter on the use of digital communication technology for development activities in Kenya. The authors devote chapter 12 to mobile and web-based technologies. They argue that these digital communication technologies distribute power among different players instead of sustaining the unequal power distribution in the top-down traditional approach. They show how mobile technologies have improved banking via a system called M-PESA, animal husbandry (iCOW) and health (mHealth). They discuss the world's first mobile phone cow calendar and how the iCow has been effectively used by farmers. To do this they produce several narratives from farmers. They provide similar case examples in development sectors focusing on health and banking. A key point made by the authors is that digital technology has been critical to participation of low income earners in their own self-sufficiency.

Opata's contribution in chapter 13 is the final country case and it analyzes possibilities of telecommunications services for development in rural Nigeria. He presents a business case for telecommunications services in areas where many scholars have argued that cost issues make it difficult for such services to be sustainable. Opata uses both macro- and microeconomic studies to show effects of telecommunications access to a nation's economic growth and to individual lives in rural areas respectively. This, he argues, is sustainable because rural access will advance social inclusion and environmental protection. He demonstrates, with several examples, how individual lives in rural areas are improved with access to telecommunications. Though he acknowledges that service in rural area is more likely to occur via voluntary corporate social responsibility, he notes that such voluntarism could also become profitable.

This book is one that we had conceptualized about a year ago and considered important largely because of the continued important to talk about and repeat such talks or as the title of the book informs us—re-imagine—development communications in a new era of new technologies, new types of governance, among other continued changes in our world and continent. As we noted above, several of the chapters have largely captured development communication in the lights of our changing world. We thank Daniel Awodiya, Annie Duru, Matt Mogekwu, Ritchard M'Bayo, Cornelius Pratt, Thomas Jacobson, Genet Lakew, Hilary Ozoh, Tokunbo Ojo, Jan Servaes, Kehbuma Langmia, Eddah Mutua-Kombo, Ayantu Tibeso, George Gathigi, Ernest Waititu, and Bede Opata for their contributions to this book. Finally, we express gratitude to the scholars whose research and contributions to the field of development communication have enlightened us over the years. To them we owe an eternal gratitude.

Part 1

Reviewing Past Frameworks and Today's Discourse

Chapter 1

Transcultural Social Change and Development: A Critical Review of Paradigm Shifts

Daniel Awodiya

The term *development* has been viewed from many different academic and pragmatic perspectives. *The Economist* had once equated increase in per capita income and annual growth rate to development, thus discounting social transformation. The dependency theorists viewed development from the perspective of the rich nations dominating the underdeveloped nations. By this power relationship, it was thought that there would be a transfer of some measure of development to the weaker nations. The United Nations, which is the umbrella organization of the nations of the world, recognized the importance of social transformation and thought of development as the sum of the equation of economic growth and social change.

Other scholars believed that for development to occur there must be fundamental changes in individual skills and capacity, level of freedom, self-discipline, responsibility, and material well-being. The totality of these changes, by implication, will reflect the state of the society or a nation as a whole. Yet, to some, development is concerned with the eradication of poverty, illiteracy, preventable diseases, unemployment, and inequality.

The humanists focused on the human and ethical dimensions of development. From this perspective, development is seen as liberation from poverty and from a stunted view of self. And, in the same vein, it is believed that merely increasing consumption or expanding material benefits for the individual is insufficient and in the end dehumanizing.[1] In sum, according to Simon, the term *development* encompasses strategies and programs to improve the living conditions of target populations.[2]

Who determines the innovation? What type of change? Change from and for what? And change in whose interest and at what cost? These questions raise critical issues. As Opubor again, notes: any notion of development in Africa, for example, that does not take into account fundamental change, for better, in the lifestyle of peasant farmer, the fisherman, the petty trader, the herdsman, or laborer, is grossly inadequate.[3] It should be added that while development efforts should accomplish all of these goals, it must not be at the cost of stripping the locals of their culture, dignity, and humanity.

What is development? Is it economic growth or science-based technological advancement? Is it modernization of institutions, including education philosophies, and is it political empowerment and sovereignty? Following is an overview of the main development paradigms, their overarching goals, limitations, and examples.

Traditional Perspectives

The social interventionist traditionally relied upon the blueprint of innovation diffusion by Rogers and Shoemaker, the classical mass communication works of Lerner, Klapper, Pye, Schramm, and others.[4] Theoreticians in this paradigm view developing nations as inferior and touted ideas that were totally foreign to the social-cultural realities of the people. Nwuneli concurs that although they had good intentions, many organizations failed because of misguided and defective theories of social transformation and development. Almost all of these social scientists argued for the transplant of some form of western-styled social development model to Africa.[5]

Nwuneli also evokes Rostow's five-step model for economic growth and development. Even the countries that matched the model's variables for development (Ivory Coast, Nigeria, and Kenya), Nwuneli notes, are far from realizing the expected economic take-off.[6] McCelland developed another model based on the same principle of alterity, evoking the idea that lack of achievement motivation is at the root of all ills of developing nations.[7] He believes, mistakenly, that people in the developing countries should mirror the achievement levels of those in the developed nations. Lerner similarly believes that people of the developing nations should empathize with those of the developed nations who enjoy higher political, social, cultural, and economic levels. This, ostensibly, will lead to the modern transformation of the developing nations.[8] It is illogical and erroneous that empathy is limited to the developed nations because empathy is multidimensional and a function of individual socio-economic and life ambitions as influenced by culture.

Finally, ownership of mass media facilities by various nations is a measure of the social development levels of those countries.[9] (Schramm, 1964). Individual modernity is linked to access to radio, television, films, telephone, newspapers, among others. However, contemporary research does not confirm the correlation of acquisition and access to media facilities with social development.[10]

Today, Western scientists that propounded and propagated these paradigms have repudiated many of the development theories of the sixties due to their failure to consider the social and cross-cultural limitations free of non-Western assumptions.[11] Nwuneli laments that this paradigm shift was at a colossal cost to developing countries that invested billions of dollars and most of their foreign aid on development and modernization projects based on incongruous Western theories and practices.[12]

Example and Critique

An example of this misplaced development approach is captured in the sentiment that was echoed by Gibbs while evaluating the work of Alec Dickson in the 1950s on mass education in Togoland.[13] He observed, correctly, that throughout the first half of the twentieth century colonial regimes across the world actively used various types of media in reinforcing their control and encouraging adoption of new practices. Inherently, Western media content was propagated for consumption by indigenous people. The recognition of theatre as a vehicle for disseminating innovation led to the infusion of development messages into drama production. But Bhatia reviewed the practice and concludes that much use of theatre and dance in development, concentrated mainly on the dissemination of messages chosen by those in power to encourage people to change their behavior patterns.[14] The plausible criticism by Paulo Friere and others prompted modification of unidirectional dissemination of such ideas that model Western agents of development and instead suggested integrated approach that encourages input of local ideas and ingenuity into solving local problems.[15]

The Social Marketing Paradigm

Social marketing, another popular Western-oriented theoretical formulation, is the application of marketing and promotional strategies to sell identified social ideals and behaviors and to induce attitude shift and behavioral alignment on the part of targeted populations. Wiebe's proposition compelled not-for-profit practitioners to consider applying commercial marketing principles to achieve behavioral changes in their respondents.[16] The aim of this paradigm is to rid the world of its malaise of illiteracy, poverty, overpopulation, disease, conflict, underdevelopment, and the like. Now, researchers and practitioners long concerned with selling products are turning their attention to selling healthy, safe, and socially positive behaviors.[17] But the question is "can brotherhood be sold like soap?"

SM incorporates careful product development, a consumer-oriented perspective, segmented targets and programs, and a systemic process of analysis, planning, implementation, and re-planning.[18] The marketing-oriented social programs utilize the full application of essential marketing elements such as offering product, price strategies, channel strategies, distribution strategies, and communication. Social marketing is advantageous because it offers an emphasis on research, an interdisciplinary perspective, a communication mix model, a theory of exchange, and a client orientation.[19]

The social marketing paradigm has been widely accepted in the developed countries, especially in the West, and has been exported to the developing world and adapted to myriad of social causes. For example, Novelli notes that contraceptive social marketing (CSM) programs in developing countries have been going on now for several years.[20] These programs, funded by the U.S. Agency for International Development (USAID), are active in Jamaica, Colombia, Bangladesh, Nepal, Mexico, and many other nations.

Many African social mobilization agents have become disenchanted with the SM paradigm such that they are criticizing what they call the over-marketing of social marketing. Czaplicki declares that considering health and education as commodities to be sold by persuasion and communicational messages of educational television spots, radio jingles supported by interpersonal communication of extentionists, is an arrogant conception disseminated in developing countries by expert communication consultants of the USAID.[21] The application of marketing approaches used in selling soda and detergents to promote concepts like family planning, disease prevention, and literacy is beset with challenges. Then, to promote cross-cultural and cross-national application of such approaches portends even greater, critical hurdles.

Example and Critique of Social Marketing Practice

Pfizer, a pharmaceutical company, which started a project called *Mobilize Against Malaria,* a five-year (2007- 2011) malaria treatment and education initiative in Kenya, Ghana and Senegal, to help address critical gaps in malaria treatment and education, offers a vivid example. Population Services International (PSI), a social marketing service provider that has experience working in these regions, was selected to promote symptom-recognition and treatment-seeking behaviors at the household level, with an emphasis on pregnant women and children under five. PSI is involved in the social marketing of insecticide treated nets (ITNs) in many parts of Africa. Today, PSI distributes nets in Angola, Democratic Republic of Congo (DRC), Ethiopia, Madagascar, Namibia, Rwanda, Tanzania and Uganda.

In evaluating the application of social marketing by PSI in sub-Saharan Africa, Omona questions if social marketing is an appropriate strategy for Africa, submitting two aspects of social marketing currently adopted by PSI.[22] He claims that they are inappropriate to effectively address the malaria pandemic in the continent. The concerns are that of charging a price for the ITNs and focusing on a segment of the community such as the children under five and pregnant women only, ignoring the bulk of the community. By implication, he argues, that the strategy of PSI suggests that if one is infected with malaria but is not under five years and not pregnant, he/she would not benefit from ITN. It further suggests that if one can even be infected, but if he/she cannot afford the subsidized ITN, he/she misses on the service. Considering the Abuja declaration (which promotes access and equitable treatment of people with health needs in Africa) and other social policies that enhance access and promote equity in this regard, social marketing in this case appears to operate contrary to these criteria.

The Cultural Imperative Perspective

It is argued that social science is a form of imperialism. This perception questions the universalism of social science, and by implication, the validity of cross-cultural application of social development and social change theories. Social science is different from natural science and the difference lies between natural phenomenon and the social case. Parsons confirms that human beings, their actions and cultural achievements, are embodiments of value toward which we must, in some degree, take a value attitude.[23] Value system are diverse; there is a plurality of different possible systems. Therefore, there can be no universally valid system of general theory in the social sciences. It must be emphasized, however, that this does not subvert the science in social science—it does not become mere manifestations of sentiments. Though a value element enters into the selection of the material of science, once this material is given, it is possible to come to objectively valid conclusions about the causes and consequences of given phenomena, free of value judgment and, hence, binding on anyone who wishes to attain the truth, regardless of what other subjective values he may hold.[24] In the face of critiques against the universal application of Western positivistic social scientific paradigms of empirical generalizations, therefore, a cultural reflection approach to explaining social phenomena is imperative. This discourse according to Carey is critical, radical, and non-positivistic.[25]

There is substance, therefore, in the Asante dualism of separating Eurocentrism from Afrocentrism. Both experiences are culturally valid, thus, constituting unique social actions whose distinction is cultural. Asante decries the blanket application of what he labels provincial scholarship across cultural boundaries without cognizance of the cultural phenomenon.[26] This phenomenon does not question the issue of human beings being biologically same, but it confirms radical differences in human realities, which is a consequence of cultural mythos that moderate people's behaviors. And when such people define their situation and experiences as real, they have real and unique consequences. For example, unlike Western societies, African families are closely knit and the emphases on both immediate and extended family systems and their relationships are prevalent, despite intense contacts with the West.[27] According to Nwankwo and Nzelibe, the pervasive influence of the family unit and its connections with village, town, and nation provides the African with a totality of social experience, which is stabilizing.[28]

The harmonious relationship and connection of the African with other institutions of society may prove impenetrable and, sometimes indecipherable to the outside thinker who is armed with social change and development theories. The fundamental position of these change agents is that change progresses from the level of the individual to that of the society. That is, the unit of analysis for most social change efforts is the individual rather than the society or other subsets of the society. This Western, individualistic construction, as with many others, does not mirror the realities of the African society. For example, if the cultural group collectively resists the expected social change, the focus on the individual is futile.

Social Mobilization Paradigm

Social mobilization as a social change paradigm uses social communication to create awareness, social conscience and sensibility, as well as capacity in the people themselves to resolve their own problems. According to Czaplicki, social mobilization could be referred to as process of democratic social interaction under a condition of free and equal access, dialogue, and participation, in order to modify the behavior of a large number of people, to demand satisfaction of their needs through self-reliance.[29]

UNICEF describes social mobilization as a multifaceted, dynamic process involving the diffusion of information and the sharing of experiences with various societal forces in order to stimulate, educate, and involve them in a specific social goal. In order to develop concrete social mobilization strategy/campaign, UNICEF developed a model (SNAKE) of summarized different steps to be taken and social mobilization actions to be implemented.[30]

The social mobilization model has been greatly successful in developing countries such as Bolivia, Colombia, El Salvador, Burkina Faso, Syria, Senegal, Nigeria, Egypt, and Tanzania. Much of the success of the SM model has been credited to its nature and process. The SM model is radical, pervasive, goal-oriented, practical, politicizing, conscientizing, and, thus, liberating. This strategy involves all of society, including those who may stand to lose power, position, prestige, and status as a result of the desired change; it recognizes that change is a competitive, often hostile activity.[31]

Lindblom's analysis of initiating change summarizes the shortcomings of the traditional perspective of social change, positing that the success by which societal change can be initiated depends on cognitive ability of citizens. He observes, however, that current communication patterns allow only a unidirectional flow of ideas from top to down, where, only a few talk to the many. This, he asserts, cripples citizens' ability to think, as there is no exchange of ideas that would otherwise allow them to explore ideas of change.

Many scholars from the developing world such as Friere, Nwuneli, Ugboajah, Opubor, Nwankwo and Nzelibe questioned the validity of Western social change theories and paradigms on several bases, including Lindblom's.[32] And it is no accident that they all proposed critical education of the populace through community-based programs of action derived from the community and for the community.

The Indigenous Knowledge Paradigm

In rethinking approaches to social change and development, scholars have realized the pertinence of identifying and classifying indigenous knowledge of the targeted population about the object or focus of change. Pioneered by late anthropologist, Michael Warren, of the Center for Indigenous Knowledge for Agriculture and Rural Development (CIKARD) at Iowa State University, indigenous knowledge system (IKS) is a multidisciplinary body of knowledge originally applied to health, nutrition and agriculture. This paradigm evaluates knowledge and ideas of the target population of social change and couples them with scientific knowledge in subsequent development efforts. The underlying rationale is that local populations have valid experiences and practices that are better incorporated in the total development effort.

IK research looks into, among others, issues such as indigenous seed storage systems, conflicts and complementarities between IK and agricultural research, the role of ethnobotany in documentation of indigenous knowledge, medicinal plant research, family planning practices, and participatory approaches in local government planning. Professor Warren was noted for demonstrating rational health promotion for African ethnomedicine, particularly that of Yoruba knowledge and beliefs regarding health.

Many IK centers have opened throughout the developed and developing countries with the objective of categorizing and systematizing indigenous knowledge systems. Such organizations as the World Bank and UNESCO are particularly interested in IKS. As the financier of many development projects, the World Bank sees IK as a valuable resource that can be used to improve the impact of development assistance. The Bank is interested in learning from IK resource centers what the centers have gained in identifying, validating, recording, exchanging, and disseminating indigenous knowledge and practices.

UNESCO on other hand, is collaborating with the Center for International Research and Advisory Network (CIRAN) in compiling best indigenous practices for input into its Management of Social Transformation (MOST) database. CIRAN, it is noted, is a partner in the IK initiative headed by the World Bank.[33]

Emadi confirms that the wisdom and viewpoint of IK is already affecting thinking and practice in research and extension.[34] There is a growing tendency toward participatory methods in research, and toward needs-based extension activities for rural development. The mix of IK and scientific knowledge (SK) ensures a two-way exchange between theoreticians and practitioners.

The IK initiative ensures that local populations' ideas and practices are not marginalized in the process of their social transformation. The institutionalization of the interactive multidirectional model of communication ensures dialogue, cooperation, respect, and the sharing of initiatives between agents and recipients of change.

Concerns

The concern with the indigenous knowledge paradigm is not in the validity of its claims, but the proprietary and use of the knowledge so deduced. How willing are indigenous people to use and apply the knowledge distilled from the IK initiative? The Westernized African elite, who are in position to ensure the development, classification and promotion for use of indigenous knowledge, are first to condemn them as inferior. On the other hand, when indigenous knowledge is distilled and packaged as marketable knowledge, intellectuals are the first to present it, without due compensation and copyright, to the global arena where the developed economies tap into them and in turn repackage and claim them as their own. This ongoing relationship of dependency or superior and inferior relationship is so endemic that the source of indigenous knowledge development or initiative becomes imperative in order for it to achieve wide acceptance.

In a poignant article titled "*Who Steals Indigenous Knowledge?,*" Barsh, reporting the proceedings of the Annual Meeting of the American Society of International Law, clearly identifies the culpability of the intellectual class in the developing world, who attend international conferences to report findings of their research in indigenous practices in medicine, health, chemicals, and agro-allied and other areas.[35] In an analysis of papers presented and articles published in academic journals, Barsh statistically shows how widespread the problem is and advises that developing countries stem the flight of indigenous knowledge through activities of local universities and institutions that release pertinent IK without patent or compensation.

The conclusion is that while the cultivation and infusion of indigenous knowledge is necessary for a people's development, those who produce the knowledge for consumption should be aware of knowledge piracy and poaching that rob developing countries of financial and developmental rewards.

The NGO Development Paradigm

The presence of NGOs in many developing countries acting as development agents is as a result of failure of host countries to provide basic, social, medical and physical infrastructures for their people. NGOs find a good fit for their agenda in these poor countries and engage people directly, outside of government, to promote healthy living, family control, medical interventions, social reorganizations and the like, all geared toward the collective development of people and their governments.

The impasse of development in developing countries has created the need for an "outsider" to mediate in affairs of those countries. The impasse has been described as encouraging a willing international agency with more than development agenda to provide for government what government is lacking or unwilling to provide for its people, while acting as a referee between the people and the NGOs. Some would argue that the referee in this situation is often unaware or unwilling to recognize the rules of the game. Critics have argued that although the objective of various development agencies is to reduce poverty and foster development for marginal populations, the effectiveness of their mission is ultimately dependent on the underlying framework of neoliberalism. Essentially, the global economic system of neoliberalism is not equipped as an all-inclusive system that seeks to bridge social-economic inequality but is instead a biased system, which Porter and Craig describe as genetically oriented to protecting existing property rights and providing for their expansion.[36]

Kihika also writes that given the prejudicial nature of the economic system, it is difficult to see the work of NGOs as anything more than temporary and ineffective band aid solutions to Africa's recurring nightmare of poverty and underdevelopment.[37] She claims that administered by bilateral organizations, NGOs, community-based organizations (CBOs), and religious groups are now funded as private social welfare organizations intended to provide social welfare. Although these policies are targeted at social welfare provision and poverty alleviation, they are specifically implemented from an economic standpoint. The neo-liberal biases of the funding agencies notwithstanding, NGOs have come to be seen as panacea to many ills besetting African nations.

With the call for a rethinking of the dominant paradigms of development and the promotion of bottom-up integrated development ideas, the NGOs act as intermediaries between government and people, as they are strategically positioned to facilitate development process through linking both the top and bottom. Foster believes that this alliance should be seen as complementary, informing, challenging, and reinforcing one another.[38] The relationship of host and sponsor countries does not have to be adversarial because NGOs provide human resource development, social structure, social institution building and serve as instrument of social cooperation.[39]

Example

A vanguard of indigenous NGO (INGO) to NGO training is Living Earth Foundation (LEF), an environment NGO based in London, which is collaborating with a partner NGO, Living Earth Nigeria Foundation (LENF), which is responsible for implementing two rural community-based projects in Bayelsa and Cross River States, both in Nigeria. The Bayelsa project was funded by Shell, while the other was funded by the UK government Department for International Development and EC.[40] The stated project objectives of LEF include a commitment to build the capacity of local NGOs and community-based organizations (CBOs) to implement similar projects. The courses which were designed to support all learners to succeed, whether from a rural village group or a team of professional staff, include Intermediate and Advanced Qualifications in Community Development Work, Community Skills for Community Research, and Village and Community Appraisals.

The critical evaluation of the philosophy behind the education intervention of the INGOs, hinges on two questions: Who determines content and for whose benefit? The practice of African technocrats going overseas to the West for capacity training workshops at prohibitive costs is now preempted by trainers coming to the setting of the recipient to conduct such capacity training. This has three major advantages. One, the cost of travelling overseas for training is reduced if not totally eliminated, and two, the trainers would be more sensitized to local conditions and encouraged to adapt to such intervening variables that may hinder application of their training contents. The third advantage is the multiplier effect of training locals who will in turn train other locals, thus spreading the much needed expert knowledge.

The ICT4D and Globalization Paradigm

ICT is an umbrella term for that which includes any communication device or application, encompassing radio, television, mobile phones, computers, and network hardware and software, and satellite systems as well as various services and applications associated with them such as video conferencing, and distance learning.[41] The term ICT4D was coined by Kleine and Unwin to include all new technologies of communication believed to speed up development in the Third World or even "leapfrog" countries to modernization and development, avoiding different stages of development and culminating in new information and technology economy.

This is reminiscent of the media technology paradigm that equated investment in and acquisition of media technology to modernization and development. But the new wave of interactive technological innovations of the twentieth century has necessitated a rethinking of the influence of information and communication technologies (ICT) on modernization, globalization, and development. Feinson states that many development scholars have equally argued that successful economic and industrial development is intimately linked to a nation's capacity to acquire, absorb, and disseminate modern technologies.[42] Furthermore, the UN Millennium Project posits that the long-term driving force for modern economic growth has been science-based technological advance.[43] Recognizing and supporting this premise, Gordon Brown, commenting on the Smith Institute's report, identified the bedrock of economic growth as sustained investment, innovation, education, skills, science and technology.[44]

Castells accurately observes that ICT as a technology on its own will not be able to solve social problems, but its availability and wide application in society is the prerequisite for economic and social development in our world.[45] He adds that "social development today is determined by the ability to establish a synergistic interaction between technological innovation and human values. This leads to a new set of organizations and institutions that create positive feedback loops between productivity, flexibility, solidarity, safety, participation and accountability, in a new model of development that could be socially and environmentally sustainable."[46]

In the face of optimistic claims by technocrats and development practitioners and informed observers, especially at the UN that ICT is a panacea for development, Kleine and Unwin with qualified optimism conclude that "undeniably, ICTs offer a new channel for economic growth which may allow developing countries to catch the development train faster and perhaps ensure a more sustainable ride. But the declaration of the United Nations on information has not reduced the widening digital divide between the developed and the developing nations."[47] Castells reiterates this problem when he claimed that there is marked divide in the use of ICT between the industrialized world and poor countries and that diffusion of ICT is extremely uneven, with most of Africa and many other regions of the world being left in a technological apartheid. Omi nously he warned that "not all developing countries are able to connect with the new global economic dynamics. Either the south (or portions of it) enters the democratic-technological-scientific race, invests heavily in R and D, and endures the "information economy" metamorphosis, or it becomes unimportant, unexploited or unexploitable."[48]

Castells forecasts the emergence of a Fourth World and claims that "within the framework of new informational economy, a significant part of the world population is shifting from a structural position of exploitation to a structural position of irrelevance—they will not even be considered worth of exploitation, they will become inconsequential, of no interest to the developed globalized economy."[49]

That ICT is entwined with development is incontrovertible and that it is also the driving force of neoliberal policy of economic globalization is equally undeniable. Countries of the South must join the bandwagon or truly risk being left behind and relegated to obsolescence in the fast-paced technology-driven global development.

Conclusion

Theorizing about social change has come full circle. We cannot identify, for reasons of conclusive knowledge, that one theory of social change fits all. Ultimately, we must acquiesce in the inescapable truth that there is, in the least, some relevance in each of the paradigms examined in this paper.

The complexities of society and the normative nature of social change issues invalidate a universal approach to moving people on a continuum from traditional to modern. Again, every indigenous population has its age-old practices of health, economics, agriculture, medicine, among others and to ignore them in the process of advancing the people is to proceed on a journey without a map. What achieves best result must take precedence over that which is assumed to be theoretically generalizable and that which truly liberates and empowers must be valued over that which further consumption and dependency.

Participatory or integrated social change through social mobilization is valued over social marketing; better still, recognizing and furthering systematized indigenous practices and knowledge is emancipating and would have enduring effects through application. But the intellectuals of African extraction should be mindful of distilling their indigenous knowledge for the developed world without adequate proprietary and patent negotiations.

International and local NGOs have become the driving force of knowledge and social-economic interventions across the developing world and to better harness their potentials, there must be genuine collaboration between exogenous and endogenous practices to ensure top-down and down-up symbiosis.

Technological advancement and its infusion to development programs are inevitable in this age of globalism. For the digital divide to be bridged, governments (both national and international) must realize that we cannot create a Fourth World that Castells claims is unexploitable and irrelevant because a conscious, enduring effort has not been made to create a truly global village where, as the United Nations claims, everyone should have access to information technology and development as a right.

As long as social change is a choice—what type of change, in what direction, for what purpose?—becomes the prerogative of the society that desires that change. And no one nation with technological and knowledge preeminence, should scheme to control or stymie another for economic, political, and military exploitation. As we are all developing as part of the global process, as Nederreen Pieterse asserts, we should view globalization as the necessary successor paradigm to dependency and hegemony.[50] With a new mindset of critical globalism, rather than be suspect of efforts of advanced nations at development, we should all applaud and celebrate our interdependency.

There are some good in every paradigm discussed here because research and history have taught us that there are multiple pathways to development. The concern of development scholars and practitioners alike now should be to continue the ferment in the field, till we ultimately achieve that with which we have been preoccupied for decades—the real development of the developing countries.

Notes

1. Bryant and White, *Managing Development*. Boulder, CO: Westview, 1982.

2. Simon, "Development Reconsidered." *Geografiska Annaler* 79 (1997): 183-201.

3. Opubor, "Mass Communication and Modern Development." In *Mass Communication in Nigeria*, ed. Nwuneli, 183–199. Enugu, Nigeria: Fourth Dimension.

4. Lerner, *Passing of Traditional Societies*. NY: Free Press, 1958; Klapper, *Effects of Mass Communication*. Glencoe, Ill: Free Press, 1960; Pye, *Communication and Political Development*. NJ: Prentice-Hall, 1963; and Schramm, *Mass Media and National Development*. Stanford, CA: Stanford University Press, 1964.

5. Nwuneli, "Study of Audience Listeners." Paper presented at the African Studies Association Conference, Philadelphia, 1972: 201.

6. Nwuneli, "Study of Audience Listeners."

7. McClelland, *Achieving Society*. NY: Van Nostrand, 1961.

8. Lerner, *Passing of Traditional*.

9. Schramm, *Mass Media and National Development*.

10. Lerner, "International Cooperation and Communication." In *Communication and Change in Developing Countries*, ed. Lerner and Schramm. Honolu-

lu: University of Hawaii East-West Center Press, 1967; Nwuneli, "Study of Audience Listeners"; Nwuneli, "Communication and Social Development." In *Mass Communication in Nigeria*, ed. Nwuneli, 201–219. Enugu, Nigeria: Fourth Dimension, 1985; and Barney, "Mass Media Roles," *Gazette* 19 (1973): 222–238.

11. Golding, "Media Role in National Development," *Journal of Communication* 24 (1974): 39–53.

12. Nwuneli, "Communication and Social Development."

13. Gibbs, "Propaganda and Mass Education." In *African Theatre in Development*, ed. Banaham, Gibbs, and Osofisan, 13–23. Oxford: James Currey, 1999.

14. Bhatia, *Acts of Authority/Acts of Resistance*. Ann Arbor, MI: University of Michigan Press, 2004.

15. Freire, *Pedagogy of the Oppressed*. NY: Continuum, 1970 and Salhi, *African Theatre for Development*. Exeter: Intellect, 1998.

16. Wiebe, "Merchandising Commodities and Citizenship," *Public Opinion Quarterly* 15 (1951): 679.

17. Harris, R. *A Cognitive Psychology of Mass Communication*. NY: Routledge, 2009.

18. Novelli, "Nonbusiness or Social Marketing." In *Handbook of Modern Marketing*, ed. Buell, 1–12. NY: McGraw-Hill, 1986.

19. Sego, "Beyond the Four Ps." Paper presented at the 43rd Annual Meeting of the International Communication Association in Washington, DC, 1993.

20. Novelli, "Nonbusiness or Social Marketing."

21. Czaplicki, "Communication and Social Mobilization." In *Health Communication and Development*, ed. Akinfeleye, 71–87. Ibadan, Nigeria: Spectrum, 1989.

22. Omona, "Social Marketing and the Fight," *East African Journal of Public Health* 6 (2009): 317–325.

23. Parsons, *Structure of Social Action*. NY: Free Press, 1949.

24. Ibid.

25. Carey, "Origin of Radical Discourse," *Journal of Communication* 33 (1983): 311–313.

26. Asante, *Afrocentric Idea*. Philadelphia, PA: Temple University Press, 1987.

27. Uchendu, *Igbos of South Eastern Nigeria*. NY: Holt, Rinehart, and Winston, 1965.

28. Nwanko and Nzelibe, "Communication and Conflict," *Journal of Black Studies* 20 (1990): 253–266.

29. Czaplicki, "Communication and Social Mobilization."

30. Ibid.

31. Lindblom, "Initiating Change: Modes of Social Inquiry." *American Behavioral Scientist* 40 (1997): 264–277.

32. Freire, *Pedagogy of the Oppressed*; Nwuneli, "Communication and Social Development"; Ugboajah, *Mass Communication, Culture and Society*; Opubor, "Mass Communication and Modern Development"; and Nwanko and Nzelibe, "Communication and Conflict Management."

33. Liebenstein, "Incorporating Indigenous Knowledge." Paper presented at ILO-INDISCO Workshop, Davao city, Philippines, 1998.

34. Emadi, "Theoretical Aspects of Indigenous Knowledge." Paper presented at the Rural Resource Centre, Tehran, Iran, 1998.

35. Barsch, "Proceedings of the Annual Meeting." *American Society of International Law* 95 (2001): 153–161.

36. Porter and Craig, "Third Way and Third World." *Review of International Political Economy* 11 (2004): 387–423.

37. Kihika, "Development or Underdevelopment." *Journal of Alternative Perspectives in the Social Sciences* 1 (2009): 783–795.

38. Fosler, *Local Economic Development*. Washington, DC: International City/County Management Association, 1991.

39. Pieterse, "Development of Development Theory." *Review of International Political Economy* 3 (1996): 541–564.

40. Robert and Lillis, "Accredited Learning Frameworks." *Development in Practice* 11 (2001): 86–117.

41. Kleine and Unwin, "Technological Revolution." *International Journal of Emerging Technologies and Society* 9 (2011): 35–49.

42. Feinson, *National Innovation Systems*. Tempe, AZ: Centre for Science, Policy and Outcomes, 2003.

43. UN Millennium Project, "Innovation: Applying Knowledge." London, UK: Earthscan Publishing, 2005.

44. Smith Institute, *Going for Growth*. London: The Smith Institute, 2005.

45. Castells, *Information Technology, Globalization and Social Development*. UNRISD, 1999.

46. Castells, *Information Technology*, 1–13.

47. Kleine and Unwin, "Technological Revolution," 35–49.

48. Castells, *Information Technology*, 15–43.

49. Castells, *Information Technology*, 15–43.

50. Pieterse, "Development of Development Theory."

Bibliography

Asante, M. *The Afrocentric Idea*. Philadelphia, PA: Temple University Press, 1987.

Barney, R. "Mass Media Roles in Development: A Descriptive Study from Four Developing Areas." *Gazette* 19 (1973): 222–238.

Barsh, R. "Proceedings of the Annual Meeting." *American Society of International Law* 95 (2001): 153–161.

Bhatia, N. *Acts of Authority/Acts of Resistance: Theater and Politics on Colonial and Postcolonial India.* Ann Arbor, MI: University of Michigan Press, 2004.

Bryant, C., and White, G. *Managing Development in the Third World.* Boulder, CO: Westview Press, 1982.

Carey, J. "The Origin of Radical Discourse on Cultural Studies in the United States." *Journal of Communication* 33 (1983): 311–313.

Castells, M. *Information Technology, Globalization and Social Development.* United Nations Research Institute for Social Development (UNRISD), 1999.

Czaplicki, S. "Communication and Social Mobilization for Child Survival." In *Health Communication and Development*, ed. R. Akinfeleye, 71–87. Ibadan, Nigeria: Spectrum, 1989.

Emadi, M. "Theoretical Aspects of Indigenous Knowledge." Paper presented at the Rural Resource Centre, Tehran, Iran, 1998.

Feinson, S. *National Innovation Systems: Overview and Country Cases.* Tempe, AZ: Arizona State University Centre for Science, Policy, and Outcomes, 2003.

Fosler, S. *Local Economic Development: Strategies for a Changing Economy.* Washington, DC: International City and County Management Association, 1991.

Friere, P. *Pedagogy of the Oppressed.* NY: Continuum Publishers, 1970.

Gibbs, J. "Propaganda and Mass Education: Alec Dickson and Drama for Development in the Gold Coast." In *African Theatre in Development,* ed. Banahm, Gibbs, and Osofisan, 13–23. Oxford, UK: James Currey, 1999.

Golding, P. "Media Role in National Development: Critique of a Theoretical Orthodoxy." *Journal of Communication.* 24 (1974): 39–53.

Harris, R. *A Cognitive Psychology of Mass Communication.* NY: Routledge, 2009.

Kihika, M. "Development or Underdevelopment: The Case of Non-governmental Organizations in Neoliberal Sub-Saharan Africa." *Journal of Alternative Perspectives in the Social Sciences* 1 (2009):783–795.

Klapper, J. *The Effects of Mass Communication.* Glencoe, IL: Free Press, 1960.

Kleine, D., and Unwin, T. "Technological Revolution, Evolution and New Dependencies: What's New About ICT4D?" *International Journal of Emerging Technologies and Society* 9 (2011): 35–49.

Lerner, D. *The Passing of Traditional Societies: Modernizing the Middle East.* NY: Free Press, 1958.

———. "International Cooperation and Communication in National Development." In *Communication and Change in Developing Countries,* ed. Lerner and Schramm, 103–111. Honolulu, HI: University of Hawaii East-West Center Press, 1967.

Liebenstein, V. "Incorporating Indigenous Knowledge in Development: Present Situation and Future Prospects for Donors and Indigenous People." Paper presented at ILO-INDISCO workshop, Davao City, Philippines, 1998.

Lindblom, C. "Initiating Change: Modes of Social Inquiry." *American Behavioral Scientist* 40(1997), 264–277.

McClelland, D. *The Achieving Society*. NY: Van Nostrand, 1961.

Novelli, W. "Nonbusiness or Social Marketing." In *Handbook of Modern Marketing*, ed. Buell, 1–12. NY: McGraw-Hill, 1986.

Nwanko, R. and Nzelibe, C. "Communication and Conflict Management in African Development." *Journal of Black Studies* 20 (1990): 253–266.

Nwuneli, O. "A Study of Audience of Listeners' Choice on Radio Nigeria." Paper presented at the African Studies Association Conference, Philadelphia, 1972.

———. "Communication and Social Development in Africa." In *Mass Communication in Nigeria: A Book of Reading*, ed. Nwuneli, 201–219. Enugu, Nigeria: Fourth Dimension Publishers, 1985.

Omona, J. "Social Marketing and the Fight Against Malaria in Africa: Population Services International and Insecticide Treated Nets (ITNS)." *East African Journal of Public Health* 6 (2009): 317–325.

Opubor, A. "Mass Communication and Modern Development in Nigeria." In *Mass Communication in Nigeria: A Book of Reading*, ed. Nwuneli, 183–199. Enugu, Nigeria: Fourth Dimension Publishers, 1986.

Parsons, T. *The Structure of Social Action*. NY: The Free Press, 1949.

Pieterse, J. "The Development of Development Theory: Toward Critical Globalism." *Review of International Political Economy* 3 (1996): 541–564.

Porter, D. and Craig, D. "The Third Way and the Third World: Poverty Reduction and Social Inclusion in the Rise of 'Inclusive' Liberalism." *Review of International Political Economy* 11 (2004) 387–423.

Pye, L. *Communication and Political Development*. NJ: Prentice-Hall, 1963.

Robert, M., and Lillis, F. "Accredited Learning Frameworks and NGO Capacity Building Programmes in Nigeria." *Development in Practice* 11 (2001): 86–117

Salhi, K. *African Theatre for Development: Art for Self-Determination*. Exeter: Intellect, 1998.

Schramm, W. *Mass Media and National Development*. Stanford, CA: Stanford University Press, 1964.

Sego, T. "Beyond the Four Ps: A Theoretical Explication and Research Agenda for Social Marketing." Paper presented at the 43rd Annual Meeting of the International Communication Association, Washington, D.C, 1993.

Simon, D. "Development Reconsidered: New Directions in Development Thinking." *Geografiska Annaler, Series B, Human Geography* 79 (1997): 183–201.

The Smith Institute. *Going for Growth: Science, Technology and Innovation in Africa*. London: The Smith Institute, 2005.

Uchendu, V. *The Igbos of South Eastern Nigeria*. NY: Holt, Rinehart and Winston, 1965.

Ugboajah, F. *Mass Communication, Culture and Society in West Africa*. NY: Hans Zell Publishers, 1985.

UN Millennium Project. *Innovation: Applying Knowledge in Development, Task Force on Science, Technology, and Innovation*. London, UK: Earthscan, 2005.

Wiebe, G. "Merchandising Commodities and Citizenship on Television." *Public Opinion Quarterly* 15 (1951): 679.

Chapter 2

Africa, Development and Communication: The Day After. . .

Chuka Onwumechili and Ike Ndolo

Over the past few decades, from the 1960s through the end of the last millennium, there were debates surrounding appropriate focus of development messages and media for delivering those messages to African communities. On the issue of message focus, several scholars supporting Daniel Lerner's initial focus on economic improvements dominated earlier debates during the period of political independence in most African countries.[1] Lerner argued that newly independent states should focus on transitioning from traditional to modern states by emulating the trajectories of Western nations. In essence, such transitions would entail improvements in Gross Domestic Products (GDP's) by increasing productivity via industrialization and modernization, among other strategies. However, failure to achieve these GDP goals eventually led to frustration. Thus, by the next decade, a humanistic perspective emerged calling for a focus on people rather than economics.[2] This perspective focuses on improvements on quality of life through education and provision of services in such areas as health. But as Moemeka has pointed out, "the (economic) and the (human) aspects of life are complementary in the process of development."[3] Thus, later scholars began to combine the benefits of both the economic and humanistic paradigms, describing them in terms of *pro bono publico*.

The second track of the development debates pitted advocates of indigenous media versus modern mass media of the time. At the time that

Lerner and the earlier advocates of economic improvement and trickle down policies put forward their modernization solutions, the obvious means of communication to achieve those goals focused on the use of modern mass media, particularly broadcasting. Newspapers were not popular because of literacy issues and television was not preferred because of its limited reach in the rural areas without electricity. Thus, most of the focus was on the use of radio broadcasting since radio sets were far less expensive, could overcome literacy barriers, and operated for long periods on batteries alone. However, advocates of these modern media were soon challenged by scholars such as Ugboajah.[4] Ugboajah did not quite argue against the use of modern mass media. Instead, he pointed to the complementarity of traditional media such as songs, theater, dance, as cultural-bound media that are helpful in delivering development messages to the grassroots and villages. In a sense, he was among those who argued for the merit of using an intermixture of both types of media in so far as such intermixture assured a large reach. Others who saw the importance of such intermixture include Salawu, who argued that; ". . . both the formal (modern mass media channels) and informal (oramedia) channels could also come together to form what is referred to as an integrated approach."[5]

Since the end of the era of vigorous development communication debates, there have been several issues require us to re-assess development communication in Africa through various lenses. One, there has been a rapid introduction of new communication technologies in Africa. For instance, there has been rapid spread of wireless communication technologies fueling new ways of communication and re-invention. How do these technologies affect development? What roles do they play now or may they play in the future? Two, there have been some progress in various areas of development. What types of communication assisted in those achievements and why? What are remaining impediments? Three, the widespread democratization and liberalization of communication industry have dramatically changed or shifted since the 1990s in several African countries. How have those changes affected development communication? Urbanization has increased in recent years. What development communication effects have been associated with this increasing urbanization? These are indeed varied and perhaps broad questions. Yet they are important if we wish to take stock of changes that have taken place since the development communication debates began during early periods of widespread political independence in Africa.

In essence, the purpose of this closing chapter is to take stock of development communication in the Africa region. We begin the chapter by describing key changes that have taken place since political independence in the early 1960s. In addition, we provide specific cases to point to how some of those changes have affected development communication and what the future may bring in the continued quest for social change in Africa.

Africa after Political Independence

The debates that we have attempted to capture in the opening section of this chapter coincided with widespread political independence of African nations in 1960 and immediately after. At independence, most African states remained under developed partly because previous colonial rulers focused on exploiting African countries for raw materials and labor. For colonial rulers, development of such states was not a key goal. Thus, several newly independent nations, under indigenous leaders for the first time, embarked on development projects post-colonialism. Many countries instituted five- and six-year development plans largely focused on improvement of infrastructure and industry with the goal of rapid transformation into modern nations. In essence, these countries were seeking the type of development that Lerner had written about in his seminal 1958 book that predicted a transition from traditional societies to modern ones. These national leaders had within their control various national media instruments including newspapers and broadcasting stations which they used to inform the public about projects. In doing this, the government saw journalists as tools for furthering government's development programs. Odhiambo, supporting this notion, argued that "Most African countries explicitly stated that it was the role of the mass media to create national unity and foster development. . . . Journalists suddenly became civil servants and government spokespeople."[6]

What we discussed above was prevalent in African socialist states such as Angola, Mozambique, and Tanzania. In those states, the government and the party centralized ownership of the mass media, the media actively practiced propaganda, agitation, and mass mobilization. Tanzania, for instance, practiced what was then known as *Ujamaa* (a Swahili word for familyhood), based on community sharing of land, property, and wealth.[7]

However, not all African nations saw development communication from a similar lens. Other countries allowed privately owned media, particularly newspapers. Nevertheless, private newspapers were still expected to support development communication policies of the government. Thus, in Kenya, Senegal, and several other states, private newspapers still supported government development policies, at least publicly.

It is this government–media relationship in Africa, both in socialist and non-socialist states, which led to the idea of development media theory advanced by the likes of McQuail and Merrill.[8] The theory focuses on government control of development communication process. That theory was later criticized by some scholars.[9] M'Bayo and Onwumechili advanced six propositions for a normative theory and nine theorems for a social-scientific theory of development media. This new theory recognizes multiplicity of development sources and the imperative of culture, two key elements ignored by McQuail in the development media theory.[10]

Notwithstanding, many African nations did not achieve development as anticipated, in spite of several national development plans. In fact, Easterly and Levine point out that most African countries, which were richer than their Asian counterparts in the early 1960s, declined in productivity just a few decades later.[11] More troubling was increased poverty in most of those nations. A World Bank report showed that between 1960 and 2001 (i.e., at the end of the millennium) twenty-eight African countries had declined in their gross domestic product (GDP) per capita and as many as eleven of those countries had income levels lower than at the time of political independence.[12] The decline stemmed from minimal or no growth in goods produced compared to rapid population increases. Significantly, the World Bank report noted that fifteen of the world's weakest growth performance by countries since 1960 are in Africa and only two of the world's top twenty performers during the same period come from the continent (see Table 2.1). Over the years, reasons have been advanced for the decline. These include persistent crises (including wars), corruption, oil renting effects, nondiversification of production, low levels of literacy, and lack of free markets.[13] Several of the reasons are true of some states but there are African countries such as Botswana that bucked the trend. The question is why? Among reasons advanced for this include low scale of corruption, prudent fiscal policy, and maintenance of budget surpluses. As we noted earlier, the use of GDPs to measure development prevailed in the years shortly after independence. However, it led to frustration periodic national plans demonstrated significant failures, in many states. Thus, while such economic measures remain useful today, scholars have argued for other indices of development progress.

Table 2.1: World Top & Worst Growth Performers Using GDP Per Capita Rate from 1960-2001

Best Performers			Worst Performers		
Rank	Country	Rate	Rank	Country	Rate
1	**Botswana**	6.4	1	Guyana	0.5
2	South Korea	5.8	3	**Ivory Coast**	0.5
3	Singapore	5.6	5	**Zimbabwe**	0.3
4	China	5.6	6	**Burundi**	0.3
5	Oman	5.4	7	**Nigeria**	0.2
18	**Egypt**	3.0	8	**Rwanda**	0.2
			9	**Ghana**	- 0.1
			10	**Senegal**	- 0.2
			11	**Chad**	- 0.4
			13	**CAR**	- 0.7
			14	**Zambia**	- 1.1
			16	**Sierra Leone**	- 1.1
			17	**Madagascar**	-1.3
			18	**Niger Rep.**	-1.5
			19	**Liberia**	-3.2
			20	**Congo DR**	-3.3

*NOTE: Table adapted from World Bank, World Development Indicators, 2005

Scholars question the use of economic indices for measuring development, noting that even in countries where the GDP had grown, poverty within the society may not have alleviated. These scholars began to call for more humanistic measures as we mentioned in an earlier section of this chapter.

Scholars, focused on development communication, also began to question the sole use of modern mass media for dissemination of development messages. They, in turn, attributed development communication failures to lack of appropriate cultural media which led to poor reach for development messages, particularly in the rural areas and among the poor and illiterate. At the time, most of Africa's people, as high as 70 percent of the population, lived in rural areas. The effect was that Africans were relatively unaffected by development projects, which focused on building up the cities and unreached by most development communication messages because of media chosen by development planners. It was this chain of events that led scholars like Ugboajah to argue that prior to colonization, African communities had several media for effective dissemination of information but these had been ignored by planners.[14] Those media had remained effective during and post colonization and were widely used by locals in rural communities. Furthermore, African communities value collectivism, in-group persuasion, and are largely expressive. Attributes of indigenous media reflect those values and, thus, Ugboajah reasoned that if development planners used indigenous media, the more effective the planners would be in reaching large numbers of the audience who resided in the rural areas. These media were oral in nature and, thus, Ugboajah coined a word for them, *Oramedia*.

Development Communication and Today's Societies

Today's environment is markedly different from the early years of political independence. Significant changes have occurred in many areas but we will address a few of those changes in the areas of increased urbanization, the rise of free market economies, widespread of new communication technologies, and changes in communication policies in Africa.

Though the United Nations' data show that the rate of urbanization has declined over the years, the share of population that lives in the urban centers has risen dramatically in Africa with the urban population projected to outnumber rural population in Sub-Saharan Africa by 2030 (see Table 2.2). In fact, the UNFPA notes, particularly, that "the urban population of Africa and Asia is expected to *double* between 2000 and 2030."[15] This remarkable change in distribution of Africa's population has many implications for both development and communication. The causes of this shift in population distribution are economic as is the case in most part of the world. However, in Africa there are additional reasons that arise from famine and conflicts which often lead to massive migrations to city centers. For instance, the protracted Liberia and Sierra Leone crises led to massive refugee movement to Monrovia and Freetown, respectively, as well as to other urban centers in neighboring countries. Urban centers traditionally represent centers of development advantages compared to rural areas. For instance, services particularly in the areas of health, education, jobs, reliable food sources, and general quality of life are more abundant in urban areas. UNFPA data on Table 2.3 show that nations with larger population proportion residing in urban centers are also those that assist child births with a skilled attendant present. This health service ensures better chances for mother and baby wellness. Note that data in Table 2.3 have been selected to include only two countries from the various regions in Africa. One of selected countries ranks highest in that region in terms of urban population proportion and the other ranks lowest.

Table 2.2: Rural/Urban Share of Sub-Saharan African Population (in Millions)

Year	Rural		Urban	
	Numbers	%	Numbers	%
1950	156	89	20	11
1975	255	79	67	21
2000	426	66	219	34
2030	622	49	648	51

Source: World Bank data 2005

On the other hand, increasing migration of large number of people to the cities has created slums around urban centers without basic water supply and sanitation. In essence, urban centers which hitherto provided advantages over the rural areas are fast losing those advantages. Recent UNFPA studies note that urban areas in Africa are increasing rapidly in poverty rates compared to rural areas.[16] The report notes that in places such as Angola, Chad, Madagascar, Malawi, Mozambique, Niger Republic, Sierra Leone, and Zambia, over 50 percent of the urban population live below the poverty line. The population movements have also tended to overwhelm urban resources and have led to other problems including rapid spread of HIV/AIDS and poverty.

Beyond urbanization, other changes have included a widespread adoption of democratic governance and free market economies in many African countries. This is not to claim that such governance or market economies were absent at independence. Instead, the point here is to identify the period when a significant number of African states moved to adopt those processes. This is important because military and one-party leadership and governance had dominated the African landscape before arrival of widespread democracy.[17] Bratton and van de Walle described the arrival of widespread democracy in the following words; "In Africa (too), authoritarian regimes are under siege. . . . During 1990, citizens took to the streets of the capital cities in some fourteen African countries to express discontent with economic hardship and political repression and to demand democratic reform."[18] Those actions were similar to the recent Arab Spring of demonstrations and call for democratization that pervaded 2011 and continued into 2012. In the end, they led to democratic elections in several African states.

Table 2.3: Urban Population Data, Birth Delivery by Health Attendant & HIV Prevalence

Region/Country	% Population in Urban	Urban Growth Rate (2005-10)	% Birth w/Skilled Attendant	% HIV Prevalence Rate 15-49 yrs (M/F)
AFRICA	39	3.2	47	
EAST AFRICA	23	3.7	35	
Burundi	11	6.8	25	2.6/3.9
Mauritius	42	1.1	99	0.9/0.2
CENTRAL AF-	41	4.1	54	
RICA	26	4.4	14	3.1/3.9
Chad	85	2.2	96	6.3/9.4
Gabon	52	2.6	70	
NORTH AFRICA	85	1.9	94	n.a
Libya	43	4.2	57	1.4/1.8
Sudan	57	1.0	83	
SOUTHERN AF-				
RICA	19	1.1	55	19.5/27.0
Lesotho	59	1.0	84	15.0/22.5
South Africa	44	3.7	41	
WEST AFRICA	60	4.1	51	n.a
Liberia	17	4.4	16	0.9/1.4
Niger Republic				

*UNFPA selected data 2007

These changes in governance and market policies have tended to weaken central control of development planning and implementation that pervaded the period immediately following political independence. Though the government still maintained control over large chunks of development activities, democratization created spaces for free market and private interests reducing government involvement. Non-governmental organizations (NGOs) mushroomed with most of them involved in a variety of development activities to help build capacity. Manji denoted this civil society phenomenon stating that "NGOs today form a prominent part of the 'development machine,' a vast institutional and disciplinary nexus of official agencies, practitioners, consultants, scholars, and other miscellaneous experts producing and consuming knowledge about the 'developing world.'"[19] Over 2,500 such organizations were listed for Nigeria alone in a recent directory.[20] In essence, the significant entry of these organizations diversified decision making in the area of development and importantly, not all NGOs necessarily focus on plans set aside by governments. Some follow global planning documents, others follow plans set aside by national governments, and there are others that develop own local plans. This is a world of difference from the environment that prevailed immediately after political independence in the early 1960s.

The spectrum was further widened in the area of control over communication channels. As we pointed out earlier, most governments in Africa maintained ownership of broadcast media immediately after independence. However, with widespread democratization also came liberalization of media ownership around the same period. It would be easy to claim a direct link between political democratization, liberalization and free markets. After all, those processes often go hand in hand in the West. However, this was far from the case in Africa. Instead, the impetus for market liberalization came from international financial institutions and internal pressures and in some cases, such as Nigeria, market liberalization preceded democratization.[21] In other cases, pressures came from domino effect of market openings in Europe, Asia, Africa, and other parts of the world. Most importantly, the effect was a liberalized market regulated by independent regulatory agencies. For the first time, government providers of broadcast communication faced competition from private providers. As you can imagine, government no longer had sole control of development communication messages that were hitherto centralized. In this environment, communication messages became decentralized.

The new regulatory regime ushered in by liberalization of the communication environment not only reduced government control of development messages but it also created an environment that led to an increasing availability of new communication technologies in many African countries. Table 2.4 provides data on percentage of the population with increasing access to mobile telephones and the Internet from 2000 to 2010. The data, excerpted from the International Telecommunications Union (ITU) website and modified, focuses on only seventeen selected countries but it demonstrates remarkable progress in access to new technologies. Two countries, Kenya and Nigeria, made remarkable progress in access to both technologies within a decade. While other countries may not have made similar leaps of access, they made improvements in access. Gabon, for example, now has more than 100 percent access in mobile telephone subscriptions. This figure, over 100 percent, reflects the fact that individuals often own more than one telephone handset, a phenomenon that is widespread in many African countries. This phenomenon occurs because service may not be reliable and, thus, subscribers attempt to overcome this by subscribing to more than a single provider. Increased access to the Internet is also demonstrated by the data in Table 2.4; however, this is to a less extent because of several reasons which include limited broadband facilities and limited access to reliable power.

These new technologies, particularly wireless telephones, are ubiquitous and are widely used in mass communication such as mass invitations and news dissemination.[22] Importantly, in certain cases they are modified or reinvented in local usage as predicted by Rogers.[23] In Nigeria, for instance, the concept of *flashing* was developed by wireless telephone users who beep a call recipient expecting a callback. In essence, this allocates call-payment responsibilities to the recipient in a way to symbolize socioeconomic power relations. The increasing use of these technologies opens a new vista for development communication in the near future and already they are being put to use by some development communication planners as demonstrated in a case below.

Cases and Report of Development Progress

We do not claim that changes discussed in previous paragraphs constitute exhaustive rendition of changes that have taken place in recent times. However, we contend that changes *ubi supra* are significant, particularly in any consideration for creating space for the design of appropriate development communication messages to reach Africans. There have been some successes in achieving development goals in several African countries. In this section, we review cases from Nigeria, Uganda, and Chad.

Nigeria has reported several successes in eight key areas in the Millenium Development Goals (MDGs). It has so far reported successes in key areas such as achieving universal primary education, reducing child mortality, maternal health, and combating HIV.[24] Nigeria's MDG Report notes reduction in maternal deaths from 800 in 100,000 live births in 2003 to 545 in 100,000 births in 2008 and improvement in access to AIDS treatment from 16.7 percent in 2007 to 34.4 percent in the next year.[25] Nigeria's success in maternal health has come through the National Primary Health Care Development Agency (NPHCDA) which has increased number of skilled midwives in the rural areas and equipped them with appropriate kits. These midwives, numbering in thousands, are deployed to 652 facilities clustered around 163 hospitals. In addition, they have received support at both state and rural levels via opinion leaders, key local individuals, and ward development committees. The system provides midwives with access to video conferencing system and voice communication which keeps them connected with health experts in the urban centers. However, failures persist in other areas such as environmental sustainability where forest coverage of the country shrunk from 14.4 percent in 2000 to only 9.9 percent in 2010.

Table 2.4: ITU Data on Percentage of Population with Access to New Technologies

Country	Wireless/Cellular Telephones			Internet/Web Access		
	2000	2005	2010	2000	2005	2010
Algeria	0.28	41.4	92.42	0.49	5.84	12.50
Angola	0.19	9.77	46.69	0.11	1.14	3.13
Benin Rep.	0.85	7.81	79.94	0.23	1.27	1.40
Burkina Faso	0.21	4.46	34.66	0.08	0.47	4.00
Cameroon	0.66	12.83	44.07	0.25	1.40	1.70
Chad	0.07	2.15	23.83	0.04	0.40	0.75
Ethiopia	0.03	0.55	8.26	0.02	0.22	7.23
Gabon	9.71	53.74	106.94	1.22	4.89	9.55
Ghana	0.68	13.28	71.49	0.15	1.83	25.90
Kenya	0.41	12.95	61.63	0.32	3.10	3.86
Lesotho	1.10	12.09	45.48	0.21	2.58	28.43
Nigeria	0.02	13.29	55.10	0.06	3.55	28.43
Sierra Leone	0.29	n.a	34.09	0.12	0.22	0.26*
South Africa	18.63	71.06	100.48	5.35	7.49	12.3
Togo	1.04	8.02	40.69	1.91	4.00	5.38
Uganda	0.52	4.63	38.38	0.16	1.74	12.50
Zambia	0.97	8.28	41.62	0.19	2.85	10.00

*This is data from 2009.

In Nigeria's case, we note the use of new technologies in maintaining communication with key health providers in the rural areas. The question is why has this been successful and environmental sustainability has not. The answers lie in leadership and organizational differences in both areas. The former Director of NPHCDA, hired from a pool of Nigerian health professionals in the diaspora, arrived with intent to achieve goals which often forced him to circumvent bureaucratic traps of the Federal

Ministry of Health. This singular and committed focus has helped advance development and eased suffering for rural-based women.

The second case from Uganda is different. Unlike in Nigeria, Uganda is widely known to have reduced its high HIV infection rate through a persistent campaign that involved government, NGOs, religious and community leaders. The involvement of these multi providers is remarkable. In addition, Uganda's development media campaign involved significant radio, television, print, billboards, social marketing of condoms and self-treatment kits, leaflets, and sex education programs in schools.[26] The campaign became known for its largely coordinated message called the ABC (Abstinence, Become faithful, use a Condom). Remarkable results in Uganda included increase in condom use from only 7 percent nationwide in 1990 to 50 percent in rural areas and over 85 percent in urban areas. Additionally, the percentage of infected pregnant women attending antenatal clinics fell from 31 percent in 1993 to 14 percent in 1998.

Chad may not be as successful as Nigeria and Uganda, but it does have its own success stories. A significant population of Chad continues to suffer from hunger and poverty amid a war situation in the Eastern border of the country. The country is far from achieving the MDGs in several areas including reducing number of children deaths due to maternal and neonatal tetanus. However, it has achieved few success stories reducing malnutrition from 58 percent in 1990 to 32 percent in 2000 and percentage of people with access to drinking water has increased from only 17 percent in 2000 to 34 percent in 2006.[27]

The new environment and cases that we have described demonstrate that today's Africa is remarkably different from what the case was immediately after independence. In a sense, therefore, development communication discussions cannot continue to be bugged down by the same discussions pre-new era. For instance, why continue to focus on access in rural areas as the key for choosing a development media if the balance in population distribution is shifting or has shifted from rural to urban areas? If wireless telephone is a widely subscribed communication medium in Africa, why ignore them? If television reach is far greater than was the case several decades ago, should we continue to ignore television? However, this is not to claim that old media, including indigenous media, is ineffective today. In fact, far from it, there are reasons to continue to use such development media but such media, just like new media, must pass today's test for effectiveness in delivering communication messages to the right audience. Thus, as we move to the "Day After," we need to evaluate which development media remains effective.

The Day After . . .

It seems clear that attention should be focused on who constitutes development audience and what communication media are effective in reaching them? These questions address at least three of five major activities that Moemeka pointed out as guidelines for identifying information needs of a development project.[28] The other activities, identified by Moemeka, focused on assumptions of a centralized national plan for development. However, our focus here is not on a centralized plan but recognition that there may be multiple plans, a diffused one, all intended to achieve development via various sources of which some may be non-governmental. We have already noted an example of this with the Ugandan case presented earlier. Moreover, our analysis of that case and others in the previous section provides us with insights on probable answers to the questions and we will make those clear in this section.

As we pointed out, the national development audience today is increasingly shifting from rural to urban area, increasingly literate, and has access to new communication media. However, it is notable that there is a significant proportion of that audience that remains rural and illiterate. Though data shows that approximately 40 percent of Africans now reside in the urban areas, it also shows that this varies from country to country (see Table 2.3). For instance, while countries like Libya, Gabon, South Africa, and Liberia have at least 60 percent of their populations residing in urban centers; countries like Burundi, Lesotho, and Niger Republic have fewer than 20 percent of their citizens in urban centers. These disparities do not end with population distribution across rural and urban areas, but extend to literacy rates and access to different communication media. While we know that traditional or indigenous media are still extensive in many of Africa's rural areas, we also realize the increasing usage of modern mass media. These modern media are not accessed equally across the continent as Table 2.5 demonstrates. For instance, radio remains the most prevalent mass medium in the West and East regions of the continent in spite of increasing usage of other media. In North Africa, radio has been surpassed by television as the prevalent medium even beyond the major cities. This portends different approaches to selection of development communication media in these regions and countries. It cannot be one size fits all.

Radio, of course, remains ubiquitous in most parts of Africa and is especially critical in West and East Africa where it is the most widely owned mass medium. It is also language friendly which means that even those that are illiterate in the widely-spoken language of the country or unable to read and write in the local ethnic language are still accessible. Additionally, radio is less expensive compared to other media and is usable via battery which is particularly important when one considers that a large part of rural areas in West and East Africa do not have electrical power access. Clearly, these attributes make radio a sustainable medium for reaching a large number of people. However, radio does have drawbacks that should be acknowledged. A major one is the fact that demonstrations of an activity cannot be visualized. This is very important as an innovative or complex product, service, or idea is difficult to introduce or demonstrate solely via radio.

Among the new media, wireless telephones present important options for development communication. Table 2.4 shows that this is an increasingly accessible communication medium as more and more Africans, both in urban and rural areas, own a telephone. However, it is hindered by the limited accessibility for those who may not be able to read and write. By this we mean that the illiterate may be restricted to using the medium for voice communication and is unable to use it for written communication. In essence, while the medium provides opportunities through texting to reach a large number of people instantaneously, it is useless if the audience is unable to interpret the text. However, in spite of this problem, texting has become a popular means of disseminating messages to a large number of people in Africa. For instance, it is crucial to the success of the midwifery program developed by NPHCDA in Nigeria as we pointed out earlier. We also have come to realize the importance of this medium in building large social movements during the Arab Spring in the north region of Africa.

Television is increasingly replacing, or at the least, complementing radio as an important medium for development communication particularly in North Africa but also increasingly in Southern Africa where subscription rates have also risen. Television does not only have the attributes of radio but it eases demonstration of an innovation because the audience can visualize demonstration of an innovation. However, the downside is that television usually requires electricity, which is not available or is intermittent in several West and East African countries. Discussing a few of these modern mass media separately does not indicate a recommendation for separate development communication use. In fact, most successful development communication campaigns involve use of multimedia. Thus, communication media are best used in combination. In addition, there are other media that may be added to the mix such as billboards, leaflets, and newspapers. Furthermore, Oramedia remain important in spite of the fact that rural areas have increasingly lost their large resident population in several countries. As we pointed out earlier, there are countries where a large number of the people still reside in the rural areas and need to be connected for a development communication campaign to be successful. Additionally, a means of reaching people particularly in remote areas is through opinion leaders who are often local. These could be religious leaders, teachers, traditional rulers, and those with high socioeconomic positions within that community. Rogers has pointed out the importance of opinion leaders particularly at the stages of persuasion, decision, implementation, and confirmation of innovation adoption.[29] Rogers' work points out that these opinion leaders are also important in the earlier stages of awareness and knowledge, particularly in the rural areas.

Table 2.5: Percentage of Population Subscribing to Media Types in 2007

Country	TV	Radio	Cable TV	Satellite Dish
EAST AFRICA				
Burundi	3.0	69.0	4.0	1.0
Kenya	37.0	92.0	1.0	1.0
Tanzania	20.0	93.0	5.0	3.0
Uganda	8.0	83.0	1.0	1.0
NORTH AFRICA				
Algeria	96.0	79.0	35.0	88.0
Egypt	98.0	78.0	38.0	44.0
Morocco	99.0	81.0	1.0	85.0
Tunisia	100.0	86.0	1.0	81.0
SOUTHERN AFRICA				
Angola	96.0	85.0	17.0	22.0
Mozambique	76.0	90.0	5.0	5.0
South Africa	76.0	89.0	9.0	0.0
Zimbabwe	35.0	56.0	1.0	4.0
WEST AFRICA				
Ghana	50.0	89.0	5.0	1.0
Nigeria	53.0	90.0	8.0	2.0
Senegal	69.0	91.0	7.0	9.0
Togo	28.0	87.0	2.0	1.0

Source: Balancing Act (2008).

Conclusion

The goal of this chapter was to take stock of development and development communication in Africa from the early years of political independence to date. Therefore, we have attempted to review the history of the field noting bifurcation of ideas early on with debates between scholars who stressed importance of economic measures that were largely measurable through GDPs and other scholars who detested such measures and instead recommended a focus on human elements. The division also existed in the narrower area of development media where there existed proponents of modern mass media versus proponents of traditional or indigenous media. Ultimately, scholars sought middle ground in both debates, recommending mixtures of measures and media.

The chapter noted key changes that drive the field today. These come in the areas of increased urbanization and its effects; widespread democratization and free market policies that tend to weaken the once powerful presence of national governments in development activities; and widespread introduction and use of new communication technologies. Each of those elements have had varied but remarkable impact on development communication. These elements are noted in the cases presented in the chapter from Nigeria, Uganda, and Chad. These cases point to success in each of those countries.

Finally, we argue that changes in the African environment driven by key elements noted earlier and underlined by country cases support a focus on understanding the audience and assessing appropriate media for reaching such audience. While the chapter analyzes several media separately, it concludes by recommending that a multimedia approach is often required.

Notes

1. Lerner, *Passing of Traditional Society*. NY: Free Press, 1958; Rostow, *Stages of Economic Growth*. NY: Cambridge University Press, 1962; and Schramm, *Mass Media and National Development*. Stanford, CA: Stanford University Press, 1964.

2. Inayatullah, "Toward a Non-Western Model." In *Communication and Change*, ed. Lerner and Schramm, 98–102. Honolulu, HI: East-West Center Press, 1967; Beltran, "Rural Development and Social Development." Proceedings of Cornell-CIAT Symposium, New York State College, March 17–24, 1974; and Hornik, *Development Communication*. NY: Longman, 1988.

3. Moemeka, "Development Communication for Developing Societies." *Gazette* 59 (1997): 380.

4. Ugboajah, "Oramedia in Africa." In *Mass Communication, Culture, and Society*, ed. Ugboajah, 165–176. London: Hans Zell, 1985.

5. Salawu, "Development Communication." In *Globalization and Development*, ed. Mojaye, Oyewo, M'Bayo, and Sobowale, 13–22. Ibadan, Nigeria: Ibadan University Press, 2008, 20.

6. Odhiambo, "Development Journalism in Africa." *Africa Media Review* 5 (1991): 17–29.

7. Nyerere, *Ujamaa Nairobi*. Nairobi: Oxford University Press, 1968.

8. McQuail, *Mass Communication Theory*. Newbury Park, CA: Sage, 1987; and Merrill, *Global Journalism*. NY: Longman, 1995.

9. Yin, *Beyond the Four Theories*. Columbia, SC: AEJMC, 2008 and M'Bayo and Onwumechili, "Rethinking Development Media Theory." In *Political Culture, Cultural Universals*, ed. M'Bayo, 129–153. Lewiston, NY: Edwin Mellen Press, 2011.

10. M'Bayo and Onwumechili, "Rethinking Development Media Theory."

11. Easterly and Levine, "Africa's Growth Strategy." *Quarterly Journal of Economics* 112 (1997): 1203–1250.

12. GDP per capita is measured by adding all goods and services produced by a country divided by population. Also see World Bank, *World Development Indicators*. Washington, DC: The World Bank, 2005.

13. Oil-renting states tend to narrowly focus attention on oil receipts while ignoring production in other parts of their economy.

14. Ugboajah, "Oramedia in Africa."

15. UNFPA, *State of World Population*. NY: UNFPA, 2007, 6.

16. UNFPA, *State of World Population*.

17. Bratton and Van de Walle, "Toward Governance in Africa." In *Governance and Politics in Africa*, ed. Hyden and Bratton, 27–55. Boulder, CO: Lynne Rienner, 1992, and Onwumechili, *African Democratization and Military Coups*. Westport, CT: Praeger, 1998.

18. Bratton and Van de Walle, "Toward Governance in Africa," 27.

19. Manji, "Missionary Position: NGOs." *International Affairs* 78 (2002): 567–583.

20. Directory of Development Organizations, "Africa." www.Devdir.org

21. Onwumechili, "Nigeria: Equivocating." In *Negotiating Democracy: Media Transformations*, ed. Blankson and Murphy, 123–142. Albany, NY: State University of New York Press, 2007, and Onwumechili, "Neo-Liberal Regulation." In *Media and Communications Industries*, ed. Olorunnisola, 81–101. Lewiston, NY: Edwin Mellen Press, 2009.

22. Onwumechili and Sousa-Johnson, "Reinventing Telephone Usage." In *Political Culture, Cultural Universals*, ed. M'Bayo, 297–307. Lewiston, NY: Edwin Mellen, 2011.

23. Onwumechili and Sousa-Johnson, "Reinventing Telephone Usage," and Rogers, *Diffusion of Innovation*. NY: Free Press, 2003.

24. MDG, "MDG 2010 Report." www.mdgs.gov.ng/
25. MDG, "MDG 2010 Report."
26. Stoneburner and Low-Beer, "Population-level HIV Declines." *Science* 304 (2004): 714–718.
27. MDGMonitor, "Tracking the Millenium Development." www.mdg monitor.org.
28. Moemeka, "Development Communication for Developing Societies."
29. Rogers, *Diffusion of Innovation.*

Bibliography

Balancing Act. "African Broadcast and Film Markets." London, UK: Balancing Act (2008): www.balancingact-africa.com/node/15375.

Beltran, L. "Rural Development and Social Development: Relationships and Strategies. Communication Strategies for Rural Development," *Proceedings of the Cornell-CIAT International Symposium*, New York State College of Agriculture and Life Sciences, March 17–24, 1974.

Bratton, M., and Van de Walle, N. "Toward Governance in Africa: Popular Demands and State Responses." In *Governance and Politics in Africa*, eds. G. Hyden and M. Bratton, 27–55. Boulder, CO: Lynne Rienner Publishers, 1992.

"Directory of Development Organizations." *Africa*. Volume 1.B (2011). www.Devdir.org/files/Nigeria.PDF.

Easterly, W., and Levine, R. "Africa's Growth Tragedy: Policies and Ethnic Divisions." *Quarterly Journal of Economics,* 112 (1997): 1203–1250.

Hornik, R. *Development Communication: Information, Agriculture, and Nutrition in the Third World.* NY: Longman, 1988.

Inayatullah, I. "Toward a Non-Western Model of Development." In *Communication and Change in the Developing Countries*, ed. Lerner, D., and Schramm, W., 98–102. Honolulu, HI: East-West Center Press, 1967.

Lerner, D. *The Passing of Traditional Society*. NY: The Free Press, 1958.

Manji, F. "The Missionary Position: NGOs and Development in Africa." *International Affairs,* 78 (2002); 567–583.

M'Bayo, R. and Onwumechili, C. "Rethinking Development Media Theory through Normative Propositions and Social Scientific Theorems." In *Political Culture, Cultural Universals, and the Crisis of Identity in Africa: Essays in Ethnoglobalization,* ed. M'Bayo, R., 129--153. Lewiston, NY: Edwin Mellen Press, 2011.

McQuail, D. *Mass Communication Theory: An Introduction.* Newbury Park, CA: Sage Publications, 1987.

MDGMonitor. "Tracking the Millenium Development Goals – Chad." MDGMonitor (undated): www.mdgmonitor.org/index.cfm.

Merrill, J. *Global Journalism: Survey of International Communication.* NY: Longman, 1995.

"Millenium Development Goals." MDG 2010 Report: www. mdgs. gov.ng /mdg-report-2010.

Moemeka, A. "Development Communication for Developing Societies: Facing the Realities." *Gazette,* 59 (1997), 379–393.

Nyerere, J. *Ujamaa Nairobi.* Nairobi: Oxford University Press, 1968.

Odhiambio, L. "Development Journalism in Africa: Capitulation of the Fourth Estate."*Africa Media Review,* 5(1991), 17–29.

Onwumechili, C., and Sousa-Johnson, T. "Reinventing Telephone Usage in Nigeria." In *Political culture, cultural universals, and the crisis of identity in Africa,* ed. M'Bayo, R., 297–307. Lewiston, NY: The Edwin Mellen Press, 2011.

Onwumechili, C. Neo-liberal regulation: NITEL in murky waters. In *Media and communications industries in Nigeria,* ed. Olorunnisola, A., 81–101. Lewiston, NY: Edwin Mellen Press, 2009.

———. "Nigeria: Equivocating while Opening the Broadcast Liberalization Gates." In *Negotiating democracy: Media transformations in emerging democracies,* ed. Blankson, I., and Murphy, P., 123–142. Albany, NY: State University of New York Press, 2007.

———. *African Democratization and Military Coups.* Westport, CT: Praeger Publishers, 1998.

Rogers, E. *Diffusion of Innovation.* NY: Free Press, 2003.

Rostow, W. *The Stages of Economic Growth.* NY: Cambridge University Press, 1962.

Salawu, A. "Development Communication: The Preliminaries." In *Globalization and Development Communication in Africa*, ed. Mojaye, E., Oyewo, O., M'Bayo, R., and Sobowale, I., 13–22. Ibadan, Nigeria: Ibadan University Press, 2008.

Schramm, W. *Mass Media and National Development.* Stanford, CA: Stanford University Press, 1964.

Stoneburner, R., and Low-Beer, D. "Population-level HIV Declines and Behavioral Risk Avoidance in Uganda." *Science,* 304 (2004): 714–718.

Ugboajah, F. "Oramedia in Africa." In *Mass Communication, Culture, and Society in West Africa*, ed. Ugboajah, F., 165–176. London: Hans Zell publishers, 1985.

United Nations Population Fund (UNFPA). *State of World Population 2007: Unleashing the Potential of Urban Growth.* NY: UNFPA, 2007.

World Bank. *World Development Indicators.* Washington, DC: The World Bank, 2005.

Yin, J. *Beyond the Four Theories of the Press: A New Model for the Asian and World Press.* Columbia, SC: AEJMC, 2008.

Chapter 3

Speaking of African Women's Agency, Activism and Complementary Gender Roles: A Look at African Feminism

Annie N. Duru

The lives of African women and their social environment have received much attention in social science research particularly when it pertains to portraying African women as oppressed, helpless, and ultimately victims of patriarchal structures. The strength, resilience, and activism of these women are often overlooked as some Western scholars are quick to jump to the assumption that African women are without agency, and thus are unable to move for a social change.

However, many African feminist scholars have made counter arguments rejecting the hegemonic portrayal of African women by the West. An African scholar Ama Aidoo critiqued the portrayal of African women in Western media as victims because of their looks or how many children they have.[1] Obioma Nnaemeka in her work, *If Female Circumcision Did Not Exist, Western Feminism Would Invent It*, critiqued the West's assumptions of African women as victims, who lacked agency.[2] Also, Third World feminist, Chandra Mohanty critiqued the hegemonic ideology and discourse on African women embedded in the work of some western feminist.[3] Mohanty stresses that African women are often categorized as oppressed and powerless in Western literatures.

On the other hand, many of the literatures by African scholars highlight different ways in which African women actively participate in their social environment. Some of the literatures show that the lived experiences and status of African women are portrayed differently by the West. Quite often, they are misunderstood and misrepresented. A typical example of such misunderstanding is seen in the explanation of gender ideology in Africa. The "woman-to-woman marriage" and the "male daughters" experiences which is discussed later in the chapter show that African women are empowered by some of these experiences, because they provide women with opportunities to challenge or overcome some of the traditional stipulations that disempower them.[4] Also, the work of these African scholars show that gender ideology in Africa is complex, and in many ways are different from other parts of the world. In fact, an African scholar Oyeronke Oyewumi presents a different view of gender ideology when she suggests that "gender was not an organizing principle in Yoruba society prior to colonization by the West."[5] Her work suggests that the names and titles in Yoruba society were genderless. In other words, gender was not considered to be a priority in Yoruba culture. Oyewumi's work clearly illustrates differences in the understanding and interpretation of "gender" across cultures.

As African scholars critique portrayal of African women in Western media, they illuminate the agency of African women to pursue their interest, and advocate for a change in status quo when oppressed. This chapter analyzes African feminism to demonstrate that African women have agency and are able to participate in socio-political movements and social development. The chapter looks at the tenets of African feminism as conceptualized by Sierra Leonean anthropologist Filomena Chioma Steady, first in *The Black Woman Cross-Culturally,* and second, *African Feminism.*[6]

African Feminism

The tenets of African feminism are based on traditional African values, where gender roles are viewed as interdependent and complementary, and contribute to development and sustenance of human life. Filomina Steady describes African feminism as humanistic feminism. Steady views African feminism as an ideology that takes into account the struggles and sufferings of women, and view women as humans rather than sexual beings. African feminism embraces liberation from different forms of oppression, while bridging the gap in nationality and class among African women to offer a more inclusive type of feminism.

Steady suggests that African feminism is deeply rooted in the values of traditional African society where communalism is the norm.[7] According to Steady, men and women have to work together in traditional African society in pursuance of the well-being of the society and development. Steady also asserts that social roles are defined for African women and this gives them some sort of autonomy and independence to participate in social, economic and political activities. She uses processes of slavery, colonialism, neocolonialism, apartheid, and chronic underdevelopment to illustrate the struggles and resistance of African women.

African Women's Agency and Activism

Contrary to some Western scholars' assumptions that African women lack agency when it comes to combating oppressions against women and fighting for women's right, African feminism demonstrates the agency and activism of African women.[8] Agency refers to possessing the capability, and the means to accomplish goals. The view that African women are active players in their environment is popular among African feminist scholars. Steady suggests that African women have been engaged in activism without a feminist label attached to it. Also, Nnaemeka notes that African women are more focused on taking action to demonstrate their feminism rather than articulating it. In other words, they are more pragmatic in their feminism. In describing the framework of African feminism, Nnaemeka writes:

> The majority of African women are not hung up on articulating their feminism, they just do it. In my view, it is what they do and how they do it that provides the framework.[9]

African feminism recognizes the strength, agency, and activism of African women in resisting social, political, and economic exploitation perpetuated by colonialism. A typical example of their activism is the 1929 women's protest in Nigeria which demonstrates agency of Igbo women to protest against the colonial administration's plan to impose tax on women and men.[10] Following the imposition of tax by the colonial government, Igbo women in Aba, a city in Eastern Nigeria, organized themselves, took to the streets, and protested against high rates of tax and economic oppression of women by the colonial government. The women were discontent with men's control of the economic system. The actions of Igbo women in the protest demonstrate their activism in resisting oppression.

Similarly, Lagos Market Women's Association (LMWA), Nigeria, in 1932 protested against colonial administration's measures to levy market women.[11] In addition, the Abeokuta Women's Union (AWU) founded in the 1940s, under the leadership of Fumilayo Ransome-Kuti, demonstrated against taxation of women during the colonial era. The AWU was successful in pushing for suspension of female taxes, and in having a female representative on the transition council that will take over the government.[12]

African women's activism was significant during the struggle for independence in South Africa. Women were actively involved in the fight for independence. Women such as Sibongile Mkhabela, Winnie Mandela and many others protested and fought against oppression in a time when those who openly opposed apartheid were killed, jailed or sent on exile.[13]

Resistance to colonial oppression contributed to the emergence of African women's movement. Mikell suggests that African women recognized that their participation is important in ending colonial oppression and pursuing independence.[14] After colonial rule, African women's movement continued to grow as women experienced subjugation and oppression from men. Steady notes that there were no major changes in the social inequalities created by colonialism after independence in Africa because the economic system introduced by the colonial administration continued to exist.[15] As a result, the marginalization of women in economic and political matters continued.

Mikell notes that internal problems, such as political instabilities, must have contributed to the recognition of African feminism:

> I suspect that the greater willingness of African women to embrace feminist politics and gender representation in the 1990s is traceable to the current national crises and political transitions which have been occurring throughout the continent over the past fifteen decades.[16]

The behavior and actions of African women seem to be influenced by political climate and socio-economic factors that affect women. Additionally, post-colonialism brought a level of consciousness within African women's movement.[17] Post-colonial era meant freedom of colonial rule and control for African women. They put up resistance against oppression and marginalization of women by institutions and groups. McFadden notes that both colonialism and post-colonialism played major roles on activism of African women.[18] Further, the climate at the time raised women's consciousness and led to a shift in women's behavior. The more women experienced dominant male oppression, the more they saw the need to engage in social movements so as to oppose dominant hegemonic ideologies that perpetuated inequality. The actions of African women when oppressed show that they are strong and resilient.

Presently, more African women engage in women's movement to advocate for social change and female equality. Tripp et al. write that:

> Women's movements have successfully lobbied for constitutional reforms to include gender equity and antidiscrimination clauses.[19]

The women's movement in Africa has been successful protesting against discriminatory laws. In the 1980s in Nigeria, women NGOs protested against population control policy which affected women's reproductive rights. The population policy endorsed that a woman should only have four children, and encouraged polygyny which allows a man to have more than four children.[20] The National Council of Women Societies (NCWS) in Nigeria persuaded the government to modify the population policy.[21] In Uganda, the Ugandan Council of Women (UCW) demonstrated against the colonial administration for deporting the King of Buganda, and criticized the parties for marginalizing women leaders. Also, the UCW pushed for female literacy, encouraging women to participate in leadership so as to influence policies.[22]

Through women's non-governmental organizations (NGOs), at the international and domestic level, African women advocate for the advancement of women's status.[23] Their agency and activism is seen in their participation in global events such as the United Nations (UN) held at Mexico City (1975), Copenhagen (1980), Nairobi (1985), and Beijing (1995).[24] Tripp et al. note that number of women NGOs in Africa is large, particularly in "Kenya, Mali, Mozambique, and Tanzania," compared to other NGOs.[25]

Although there are many women-oriented NGOs in Africa that work to elevate women's status and enhance their lives, some of them are mentioned and discussed often in scholarship literature more than others. For example, the National Council of Women Societies (NCWS), an umbrella organization for women NGOs in Nigeria is known to represent and speak for Nigerian women both at the international and domestic arena, with an aim to liberate and empower Nigerian women socially, economically, and politically.[26] Also, Women in Nigeria (WIN), formed in 1982, is known for its radical ways in advocating for women's rights in Nigeria.[27] Following annulment of the 1993 presidential election in Nigeria, WIN collaborated with other human rights activists and started a media campaign to protest against human rights violations perpetuated by the military regime.[28] In their media campaign, they strongly objected against disappearance of those in opposition including politicians and pro-democracy activist.[29] WIN's actions clearly demonstrate women's activism in promoting human rights and democracy. Other women NGOs include Tanzania Media Women's Association (TAMWA) established in 1987, Uganda Media Women's Association (UMWA), and Association of Media Women in Kenya formed in 1982.[30]

In addition, African women's agency is seen in their domestic environment and in their roles as mothers. While some Western women may perceive having several children as a disadvantage or disempowerment to African women; some African women may view multiple childbirths as empowering.[31] Some scholars note that African women may use their position as mothers to question some of the demands that the culture expects them to fulfill, and in some cases women use motherhood to make demands that benefit them.[32] For example, a woman who has male children can use them to obtain land and inherit family property.

Also, the dual sex system illustrates other ways in which African women have agency. According to Amadiume, the dual sex system consists of several gender ideologies that are flexible, meaning that the gender roles and responsibility are not exactly restricted to a particular gender.[33] An example of the flexibility in the gender system is seen in the ideology of "male daughters" and "female husbands." Amadiume explained the male daughter ideology in Igbo land as a traditional ideology in which a woman is given the status of a son and she inherits her father's property, in a situation where her father did not have any surviving male child.

African women also use the woman-to-woman marriage to challenge some of the traditional restrictions placed on them.[34] This woman to woman marriage is what Amadiume explained in the gender ideology of the female husband. According to Amadiume, the "female husband" refers to a woman who initiates marriage and pays money to enter into marriage with another woman in order to produce a male child who will take on the family name and inherit the property in a traditional patrilineal pattern, or to gain access to economic resources.[35] A family where all children are females may also decide that one of the females becomes a "female husband" who marries another female with intent to reproduce a male child. This is an important decision as property inheritance in such societies is patrilineal. For instance, among Igbos of Nigeria, this provides the family opportunity to have a male child because the man who biologically fathers such a child has no claims to the child as he had not formally married the mother. Instead, the "female husband" legally claims "father" of the child through marriage. In other words, African women have agency in their domestic environment and are able to use their status as mothers or as women to maneuver cultural stipulation that discriminates against them.

Empowerment through Research

Several theories frame how researchers study and write about the experiences of women of African decent. For instance, African feminism as a theoretical framework illustrates the agency of African women. Steady suggests that African-focused gender research should apply theories that fit the context.[36] Thus, because African feminism is grounded in African perspective and reality, it is identified as a theoretical framework in studying African women. As a result, some scholars view African feminism as an opportunity to advance a framework grounded in African heritage and at the same time gives African women a voice.[37] It provides the platform for understanding conditions of existence of African women and their resistance to oppression.

Contrary to the assumptions of some Western scholars who perceive African women as victims and perhaps unable to fight against oppression, African feminism stands to give African women a voice through their participation in research. Through African feminist research, African women are given opportunity to speak for themselves and from their experiences. Participation in research gives them opportunity to contribute to scholarly knowledge. Bondi suggests that knowledge through scholarship is important since it is a source of empowerment, and that ways of learning and acquiring knowledge constitutes a concern for feminists, particularly where patriarchal conceptions affects knowledge.[38]

Rakow and Wackwitz note that feminist scholars strive to encourage a communal environment in which the voices of individuals in a society are heard and recognized.[39] Thus, African feminist research aims at empowering African women, researchers and members of their communities. African feminism demonstrates agency of women in the African continent, and their activism in fighting for women's human rights and social justice towards ending gender inequality.

Complementary Gender Roles and Social Development

Do women have significant roles in social development? To answer that question one must review African feminists' conception of gender roles. As elaborated in Steady's *The Black Woman Cross-Culturally,* African feminism does not view men as enemy because some men are also victims of oppression. Rather, it sees the oppressive forces in the society such as racism, sexism, classism, and cultural dimensions as responsible for oppressing Black men and women. It advocates for humanity. By implication, African feminism does not exclude men rather it sees African men as an ally. Steady in *African Feminism: A Worldwide Perspective* writes:

> For women, the male is not "the Other" but part of the human same. Each gender constitutes the critical half that makes the human whole. Neither sex is totally complete in itself to constitute a unit by itself. Each has and needs a complement, despite the possession of unique features of its own.[40]

Steady further suggests that within the "metaphysical" cooperation of men and women maintain "cosmological balance."[41] Male and female roles are complementary in the sense that they are interdependent in processes of production and reproduction.[42] Men and women contribute to family life, sustaining human life, and social development. In the Yoruba culture of Nigeria, it is believed that men and women's support for each other leads to the success of development in the society.[43] In other words, one gender cannot do without the other.

In pre-colonial West Africa, influence within the domestic sphere to some extent affects the authority within the public sphere, creating an overlap between the domestic and public sphere.[44] Women's contributions and reproductions in domestic sphere are important because they influence the public sphere. Thus, African women are not passive and powerless when it comes to participating in development and the economic system. Women are known for their farming and trading businesses. According to a United Nations document, African women dominate retail business, trading in food items and agricultural products, and women in West Africa dominate the trade business and are perceived to have a "history of economic independence."[45]

In addition, Ufomata, in examining economic activities of women in Nigeria, notes that women who trade in the market, usually known as market women, are economically independent and they make significant contributions in their families.[46] In some cases, women are major providers in their families. Similarly, in Uganda, women vendors known as the *toninyira mukange* significantly contribute to Uganda's economic growth following the country's economic crisis in the early 1970s.[47] According to Musisi, some women vendors support their male partners in providing for their families and some were the "sole supporters of their families."[48]

Although African women do not always get credit from western media for being economically independent and active in social development, their history and power are highlighted and recognized in African feminism. Sudarkasa emphasizes women's roles in contemporary West Africa, citing Hill:

> Throughout West Africa, most women regularly work at some occupation which directly or indirectly involves them in the cash economy. Even women in seclusions in Muslim areas carry on a variety of trade activities from which they earn cash incomes.[49]

The determination of African women to succeed is seen in their willingness to engage in social and economic activities designed to enhance their economic power and advance their livelihood. Through micro credit programs offered by international organizations such as the World Bank and United Nations, women borrow money to start up small trades, and to sustain their trade.[50] Additionally, in order to empower themselves and contribute to the daily upkeep of their family and society at large, many women participate in women's credit systems in which women pay a fixed amount on an interval and the money is collected and on a rotational basis, the money is given to a member.[51]

African women's roles demonstrate their commitment to their families and community. Their drive to embark on farming, trading, and other occupations speaks to their willingness to support their families and the community. Okonjo notes that through the rural women's credit system, women play positive roles in development of their community. For instance, Islamic African women in wife seclusion (a religious and culturally-driven ideology in which married women are required to stay within the confines of their domestic environment) also participate in socio-economic activities such as petty trading to support their families.[52] This line of argument is in concurrence with Steady's position that male and female complement each other. African women's roles speak of their strength and determination in enhancing their lives and that of their families and community.

African women also participate in politics alongside men in pre- and post-colonial Africa.[53] Okonjo observed that in pre-colonial era, women were involved in governance and decision making and at times they formed pressure groups with the aim of influencing political leaders. Okonjo's writing shows that there was a high political participation of Hausa, Yoruba, and Igbo women in Nigeria in pre-colonial era.[54] In Yoruba societies, women who had traditional titles occupied leadership positions, and governed alongside men while playing significant roles such as spiritual advisers and custodians of royal treasures. Also, Okonjo notes that in the riverine area of Igbo land, the traditional title of "Omu" is bestowed upon a woman, and by her authority, the Omu and her counselors contribute in the governance of the community while performing several social and economic functions such as overseeing the market affairs, regulating market prices, settling disputes among women, and promoting peace.[55]

African women constantly played roles in political arenas, and in most cases their presence could make a political arena entertaining and a political candidate more visible. Fall describes some of the roles of African women in political scenes: "they ensured the liveliness of political meetings, they were the ones who sat on the sidelines and applauded . . . with their colorful clothing, they gave such functions an air of festivity."[56]

Presently, the involvement of African first ladies in social, political, and economic issues in their countries illustrates complementary gender roles entrenched in the culture. The first ladies support their husbands who are presidents and heads of states, on several issues including promoting peace and stability, literacy, good health, poverty reductions, gender equality, and maintaining foreign relations.

Conclusion

African feminism provides a historical and contemporary insight of traditional African values that shape experiences of African women. It illustrates the strength, power, and resilience of African women to resist oppression and contribute to knowledge and social development. However, the rejection of feminism by some African women seems to focus on the idea that feminism is a Western ideology that excludes men and sees men as enemy. After much debate, African scholars such as Filomina Chioma Steady, Obioma Nnaemeka, Ama Ata Aidoo, recognize and acknowledge African feminism to be positive for both African women and men. Aidoo expressed that all African men and women should be feminist, if they want to contribute to African development.[57]

African feminism shows that women in Africa have agency, and are able to pursue their interest and better their lives and communities. Their activism and resilience is seen in their roles in both the domestic and public sphere. In addition, African women's involvement in social movement addresses their activism and agency. There is no doubt that African feminism provides an insight into activism, agency, and gender roles of African women.

Notes

1. Aidoo, "African Women Today," in *Sisterhood, Feminism and Power*, ed. Nnaemeka, 39–50. Trenton, NJ: Africa World Press, 1998.
2. Nnaemeka, "If Female Circumcision." In *Eye to Eye: Women Practicing Development Across Cultures*, ed. Perry and Schenck, 171–189. NY: Zed Books, 2001.
3. Mohanty, "Under Western Eyes," *Feminist Review* 30 (1988): 61–88.
4. Amadiume, *Male Daughters, Female Husbands*. London: Zed Books, 1987, and Njambi and O'Brien, "Revisiting 'Woman-Woman Marriage.'" In *African Gender Studies: A Reader*, ed. Oyewumi, 145–165. NY: Palgrave MacMillan, 2005.
5. Oyewumi, "(Re)constituting the Cosmology." In *African Gender Studies: A Reader*, ed. Oyewumi, 99–119. NY: Palgrave MacMillan, 2005.
6. Steady, "Black Woman." In *The Black Woman Cross Culturally*, ed. Steady, 7–41. Cambridge, MA: Schenkman, 1981, and Steady, "African Feminism." In *Women in Africa and the African Diaspora*, ed. Terborg-Penn, Harley, and Rushing, 3–24. Washington, D.C: Howard University Press, 1987.
7. Steady, "Investigative Framework for Gender." In *African Gender Scholarship*, ed. Bakare-Yusuf, Kisiang'ani, Lewis, Oyewumi, and Steady, 42–60. Dakar,Senegal: CODESRIA, 2004.
8. Steady, "African Feminism."

9. Nnaemeka, "Introduction: Reading the Rainbow." In *Sisterhood, Feminism and Power*, ed. Nnaemeka, 1–38. Trenton, NJ: Africa World Press, 1998, 5.

10. Ifeka-Moller, "Female Militancy and Colonial Revolt." In *Perceiving Women*, ed. Ardener, 127–157. London: Malaby Press, 1975.

11. Johnson, "Grass Roots Organizing," *African Studies Review* 25 (1982): 137–157.

12. Tripp, Casimiro, Kwesiga, and Mungwa, *African Women's Movements*. Cambridge: Cambridge University Press, 2009.

13. Aidoo, "African Women Today."

14. Mikell, "African Feminism: Toward New Politics," *Feminist Studies* 21 (1995): 405–424.

15. Steady, "African Feminism."

16. Mikell, "African Feminism: Toward New Politics."

17. McFadden, "African Feminist Perspectives," *The Black Scholar* 37 (2007): 36–42.

18. Ibid.

19. Tripp et. al., *African Women's Movements*, 1.

20. Dixon-Mueller and Germain, "Population Policy and Feminist," *Population and Development Review* 20 (1994): 197–219.

21. Tripp et al., *African Women's Movements*, 50.

22. Ibid, 44.

23. Tripp et al., *African Women's Movements*.

24. Datta and Kornberg, *Women in Developing Countries*. Boulder, CO: Lynne Rienner, 2002.

25. Tripp et. al., *African Women's Movements*, 83.

26. Ojewusi, *Speaking for Nigerian Women*. Lagos, Nigeria: All State Publishing, 1996.

27. Amadiume, *Daughters of the Goddess*. London: Zed Books, 2000.

28. Tripp et al., *African Women's Movements*.

29. Ibid, 84.

30. Tripp et al., *African Women's Movements*.

31. Nfah-Abbenyi, "Gender, Feminist Theory." In *African Gender Studies*, ed. Oyewumi, 259–278. NY: Palgrave MacMillan, 2005.

32. Emecheta, *Joys of Motherhood*. London: Flamingo, 1979, and Nfah-Abbenyi, "Gender, Feminist Theory." In *African Gender Studies*, ed. Oyewumi, 259–278. NY: Palgrave Macmillan, 2005.

33. Amadiume, *Male Daughters, Female Husbands*.

34. Njambi and O'Brien, "Revisiting 'Woman-Woman Marriage,'" and Amadiume, *Male Daughters, Female Husbands*.

35. Amadiume, *Male Daughters, Female Husbands*.

36. Steady, "Investigative Framework for Gender Research."

37. Steady, "Black Woman Cross-Culturally," and Steady, "African Feminism: Worldwide Perspective."

38. Bondi, "In Whose Words?" *Transactions of the Institute of British Geographers* 22 (1997): 245–258.

39. Rakow and Wackwitz, *Feminist Communication Theory*. Thousand Oaks, CA: Sage, 2004.

40. Steady, "Black Woman Cross-Culturally," 8.

41. Ibid.

42. Badejo, "African Feminism: Mythical and Social Power." *Research in African Literatures* 29 (1998): 94–111.

43. Ufomata, "Women in Africa," *West Africa Review* 2 (2000): 1–17.

44. Sudarkasa, "Female Employment and Family Organization." In *The Black Woman Cross-Culturally*, ed. Steady, 49–63. Cambridge: Schenkman, 1981, 52.

45. Simms, "African Woman as Entrepreneur." In *The Black Woman Cross-Culturally,* ed. Steady, 141–168. Cambridge: Schenkman, 1981.

46. Ufomata, "Women in Africa."

47. Musisi, "Baganda Women's Night Market Activities." In *Readings in Gender in Africa*, ed. Cornwall, 132–140. Bloomington, IN: Indiana University Press, 2005.

48. Musisi, "Baganda Women's," 136.

49. Sudarkasa, "Female Employment," 55.

50. Omorodion, "Rural Women's Experiences," *Journal of Asian and African Studies* 42 (2007): 479–494.

51. Okonjo, "Rural Women's Credit Systems," *Studies in Family Planning* 10 (1979): 326–331.

52. Ufomata, "Women in Africa," and VerEecke, "Muslim Women Traders," *Ethnology* 32 (1993): 217–236.

53. Okonjo, "Women's Political Participation in Nigeria." In *Black Woman Cross-Culturally*, ed. Steady, 79–103. Cambridge: Schenkman, 1981.

54. Ibid.

55. Okonjo, "Women's Political Participation," 98–99.

56. Fall, "Senegalese Women in Politics." In *African Gender Studies*, ed. Oyewumi, 233–241. New York: Palgrave MacMillan, 2005.

57. Aidoo, "African Woman Today."

Bibliography

Aidoo, A.A. "African Woman Today." In *Sisterhood, Feminism and Power: From Africa to the Diaspora*, ed. Nnaemeka, 39-50. Trenton, NJ: Africa World Press, Inc, 1998.

Amadiume, I. *Male Daughters, Female Husbands: Gender and Sex in an African Society*. London: Zed Books Limited, 1987.

———. *Daughters of the Goddess, Daughters of Imperialism: African Women Struggle for Culture, Power and Democracy*. London: Zed Books Ltd, 2000.

Badejo, D.L. "African Feminism: Mythical and Social Power of Women of African Descent." *Research in African Literature* 29 (1998): 94–111.

Bondi, L. "'In Whose Words? On Gender Identities, Knowledge and Writing Practices." *Transactions of the institute of British Geographers.* 22 (1997): 245–258.

Datta, R. and Kornberg, J. *Women in Developing Countries: Assessing Strategies for Empowerment.* Boulder, CO: Lynne Rienner Publishers, Inc., 2002.

Dixon-Mueller, R. and Germain, A. "Population Policy and Feminist Political Action in Three Developing Countries." *Population and Development Review* 20 (1994): 197–219.

Emecheta, B. *The Joys of Motherhood.* London: Flamingo, 1979.

Fall, B. "Senegalese Women in Politics: A Portrait of Two Female Leaders, Arame Diene and Thioumbe Samb, 1945–1996." In Oyewumi, O. (ed.), *African Gender Studies: A Reader,* ed. O. Oyewumi, 233–241. NY: Palgrave Macmillan, 2005.

Ifeka-Moller, C. "Female Militancy and Colonial Revolt: The Women's War of 1929, Eastern Nigeria." In S. Ardener (ed.), *Perceiving Women,* ed. S. Ardener, 127–157. London: Malaby Press, 1975.

Johnson, C. "Grass Roots Organizing: Women in Anticolonial Activity in Southwestern Nigeria." *African Studies Review* 25 (1982): 137–157.

McFadden, P. "African Feminist Perspectives of Post-Coloniality." *The Black Scholar* 37 (2007): 36–42.

Mikell, G. "African Feminism: Toward a New Politics of Representation." *Feminist Studies* 21 (1995): 405–424.

Mohanty, C.T. "Under Western Eyes: Feminist Scholarship and Colonial Discourses." *Feminist Review* 30 (1988): 61–88.

Musisi, N.B. "Baganda Women's Night Market Activities." In *Readings in Gender in Africa,* ed. A. Cornwall, 132–140. Bloomington, IN: Indiana University Press, 2005.

Nfah-Abbenyi, J.M. "Gender, Feminist Theory, and Post-Colonial (Women's) Writing." In *African Gender Studies: A Reader*, ed. O. Oyewumi, 259–278). NY: Palgrave Macmillan, 2005.

Njambi, W.N. and O'Brien, W.E. "Revisiting 'Woman-Woman Marriage': Notes on Gikuyu Women." In *African Gender Studies: A Reader,* ed. O. Oyewumi, 145–165. NY: Palgrave Macmillan, 2005.

Nnaemeka, O. "Introduction: Reading the Rainbow." In Nnaemeka, O. (ed.), *Sisterhood, Feminism and Power: From Africa to the Diaspora*, ed. O. Nnaemeka, 1–38. Trenton, NJ: Africa World Press, Inc, 1998.

———. "If Female Circumcision Did Not Exist, Western Feminism Would Invent It." In *Eye to Eye: Women Practicing Development Across Cultures*, ed. S. Perry and C. Schenck, 171–189. New York, NY: Zed Books, 2001.

Ojewusi, S. *Speaking for Nigerian Women: A History of the National Council of Women's Societies, Nigeria*. Lagos, Nigeria: All State Publishing and Printing Company Ltd, 1996.

Okonjo, K. "Rural Women's Credit Systems: A Nigerian Example." *Studies in Family Planning* 10 (1979): 326–331.

———. "Women's Political Participation in Nigeria. An Overview." In *The Black Woman Cross-Culturally*, ed. F. Steady, 79–103). Cambridge: Schenkman Publishing Company, Inc, 1981.

Omorodion, F.I. "Rural Women's Experiences of Micro-Credit Schemes in Nigeria: Case Study of Esan Women." *Journal of Asian and African Studies* 42 (2007): 479–494.

Oyewumi, O. "(Re)constituting the Cosmology and Sociocultural Institutions of Oyo Yoruba." In *African Gender Studies: A Reader*, ed. O. Oyewumi 99–119. NY: Palgrave MacMillan, 2005.

Rakow, L.F. and Wackwitz, L.A. *Feminist Communication Theory: Selections in Context*. Thousand Oaks, CA: Sage Publications, 2004.

Simms, R. "The African Woman as an Entrepreneur: Problems and Perspectives on their Roles." In *The Black Woman Cross-Culturally*, ed. F. Steady 141–168. Cambridge: Schenkman Publishing Company, Inc, 1981.

Steady, F.C. "An Investigative Framework for Gender Research in Africa in the New Millennium. In *African Gender Scholarship: Concepts, Methodologies, and Paradigm*, ed. S. Arnfred, B. Bakare-Yusuf, E. Kisiang'ani, D. Lewis, O. Oyewumi, & F. Steady 42–60. Dakar, Senegal: Council for the Development of Social Science research in Africa, 2004.

———. "African Feminism: A Worldwide Perspective." In *Women in Africa and the African Diaspora*, ed. R. Terborg-Penn, S. Harley, & A. Rushing, 3–24. Washington, DC: Howard University Press, 1987.

———. "The Black Woman Cross-Culturally: An Overview." In *The Black Woman Cross–Culturally*, ed. F. Steady 7–41. Cambridge: Schenkman Publishing Company, Inc, 1981.

Sudarkasa, N. "Female Employment and Family Organization in West Africa." In *The Black Woman Cross—Culturally*, ed. F. Steady 49-63. Cambridge: Schenkman Publishing Company, Inc, 1981.

Tripp, A.M., Casimiro, I., Kwesiga, J., and Mungwa, A. *African Women's Movements: Changing Political Landscapes*. Cambridge: Cambridge University Press, 2009.

Ufomata, T. "Women in Africa: Their Socio–Political and Economic Roles." *West Africa Review* 2 (2000): 1–17.

VerEecke, C. "Muslim Women Traders of Northern Nigeria." *Ethnology* 32 (1993):217–236.

Chapter 4

Going Back to Basics: Shifting Research Focus to Indigenous Knowledge Systems and Their Role in Development

Matt Mogekwu

Communication for development has come a long way in the last six decades or so, with the evolution of different theories and models of the relationship between communication and the development process. In general, research and practice in communication for development and social change have been guided, over the years, by models that have tended to perpetuate the emphasis on absorption of innovations that have been introduced into social systems and the change in attitude and behavior in line with these new ideas. Invariably, these new ideas tend to coincide with practices and perceptions from the industrialized world or western democracies. Mass media and communication policies have, therefore, been structured to suit and fit into the parameters established by these source countries.

For years, we have been concerned with the role of the media in the diffusion of innovations. Emphasis was continuously adjusted to accommodate issues such as the persuasion process, the top-down flow of information, economic, cultural, political and other constraints to development that are usually found in developing countries. Today, we are mostly concerned with how to adopt and adapt modern communication technologies such as the internet, cellular/mobile phones, Facebook, Twitter, YouTube and similar tools. We have concluded, more or less, that the content to be transmitted is a given and that only the mode of communication should be our focus. As a result, not much research energy is expended on alternative definitions of development and alternative sources of development ideas.

It has become necessary to draw attention to other sources of development ideas and how they can be captured by communication strategies to strengthen the development process. In particular, it is suggested here that we begin to look inwards and be guided by the likes of Ali Mazrui who has, long ago, promoted the idea of looking for, and idealizing the indigenous and promoting such as source of development.[1] It is, thus, suggested that development researchers and their communications counterparts begin to look at the whole question of Indigenous Knowledge Systems (IKS) as a viable source of development ideas and design research that will evolve communication models to help idealize such systems for development purposes. This would not amount to discarding all we know so far about development communication. It will simply add to it.

The Concept of Development (Revisited)

In the 1950s the impression was strengthening that to break out of poverty, a poor nation needed to tighten its belt, forego some consumption and invest its resources in new technologies. The locus of economic growth was thought to be industrial development

In the 1970s, it was becoming clear that for most developing countries, development actually meant "rural development," as the nature and state of the rural areas were the main issues in the development of these countries, since usually more than three-quarters of the population, productivity and income came from these areas. Uma Lele had, in *The Design of Rural Development*, designated a number of key areas of focus for development purposes in Third World countries.[2] They included agricultural extension, local farmer participation in agricultural decision-making, credit, marketing, social service, project administration and training. The point of this focus was that an improvement of the lot of the rural farmers, who were considered the backbone of most Third World countries, would translate into an overall improvement for the country.

By the early 1980s, another perspective on development had emerged. It was referred to as the Basic Human Needs (BHN) approach to development.[3] This approach focused on securing access to minimum levels of consumption of certain basic goods and services. It was an approach focused on poverty eradication within a short period as one of the main objectives of development.

Although BHN was not fully accepted worldwide as a development approach, it nonetheless had its effect on policies of bilateral and multinational donors and on developing countries. It is, however, necessary to note that much of the success of the BHN approach depended to a great extent on the availability of donor countries. But over time, these donor countries became less enthusiastic about giving. Many of them were beginning to have more domestic needs of their own. Also, it was becoming clear that aid to developing countries rarely produced the desired results. Aid never worked whether it was because the benefits were often squandered by statist regimes that preferred building empty stadia, huge cathedrals and roads to nowhere; or because a ruling elite expropriated for itself the benefits intended for the impoverished masses; or because the genuinely humane intentions of the donors were negated by western exporters who would convert aid programs into export commodity schemes.[4]

In the 1990s, an approach to participatory development started taking hold, with some benevolent organizations beginning to insist on it. Participatory development in the orientation of the Development Assistance Committee (DAC) of the Organization for Economic Cooperation and Development (OECD), is defined as a process by which people take an active hand in shaping those decisions that affect their lives.[5] The majority of those required to participate are the poor, because development is about them.

From these various perspectives on development, and there are many more, it becomes clear that development means different things to different people. But in general, as Wang and Dissanayake summarized, development is a process of social change which has as its goal improvement in the quality of life of majority of people and which makes those people masters of their own destiny by participating as closely as possible in the process.[6] Nandy noted that an important goal of development is the integration of old and new ideas, traditional and modern systems, endogenous and exogenous elements to constitute a unique blend suited to needs of a particular community.[7] According to Nandy, this approach of integrating the old and the new would consider the local culture not as something to be discarded but instead would benefit from the native wisdom in order to come up with a unique syncretization that incorporates critical traditionalism with critical modernism. This is the point where the case for indigenous knowledge systems comes in.

What Are Indigenous Knowledge Systems (IKS)?

There was knowledge before Western research. The knowledge we have acquired over the years was not all products of scientific research. We acquire knowledge everyday in various ways. Frey, Botan, Friedman, and Kreps referred to this process as "everyday ways of knowing" and identified sources such *authority, personal experience and introspection, intuition, custom*, and even, *superstition.*[8] Of course, such everyday ways of knowing have their problems, but we can still find some value in them. However, Indigenous knowledge goes far beyond everyday ways of knowing.

Many communities, groups, nations, have, over the years, developed skills, practices and from them have acquired knowledge that has been the foundation of their survival. The concept of indigenous knowledge is nothing new. Many scholars have tried to define it in various ways. According to Warren,

> Indigenous knowledge (IK) is the local knowledge—knowledge that is unique to a given culture or society. IK contrasts with the international knowledge system generated by universities, research institutions and private firms. It is the basis for local level decision making in agriculture, health care, food preparation, education, natural resource management, and a host of other activities in rural communities.[9]

Flavier et al. put it differently,

> Indigenous knowledge is (...) the information base for a society, which facilitates communication and decision-making. Indigenous information systems are dynamic, and are continually influenced by internal creativity and experimentation as well as by contact with external systems.[10]

Other definitions do not depart significantly from these two. The American Association for the Advancement of Science (AAAS) has defined traditional (indigenous) knowledge as:

> The information that people in a given community, based on experience and adaptation to a local culture and environment, have developed over time, and continue to develop. This knowledge is used to sustain the community and its culture and to maintain the genetic resources necessary for the continued survival of the community.[11]

From the foregoing, a few fundamentals can be identified and/or assumed about indigenous knowledge:

- Indigenous knowledge is not static. Just like any other element of culture, it adopts from, and adapts to, the world around it.
- It is applicable to various aspects of life such as agriculture, preventive medicine, economics, eco-diversity, use and management of nat-

ural resources, animal husbandry, savings and lending (micro-financing), and community development.

- Every human society has relied on its indigenous knowledge at a point for its survival and development.
- Indigenous knowledge can survive side by side with "modern" science.
- Today, it is a resilient part of the lives of the rural poor.

Indigenous knowledge is a reality and has played meaningful roles in development of many societies. When the different elements of indigenous knowledge come together, they form a system, the indigenous knowledge system.

A lot of work has been done to try to identify indigenous knowledge systems that have helped and are still helping different communities and societies to survive. This is true in almost every part of the world. We provide some examples below.

Barnhart and Kawagley examined the importance and salience of IKS in the education sector of Alaska natives and how this knowledge could be integrated into contemporary education theories and perspectives.[12] They identified areas of emerging research associated with indigenous knowledge systems. For example, they discussed indigenous epistemologies in which scholars have begun to identify the epistemological underpinnings and learning processes associated with IKS.[13] They looked at what they referred to as "culturally responsive pedagogy"; indigenous language learning; and place-based education, among others. It was interesting to note that in the discussion of the indigenous language learning, the following point was made that:

> Research issues associated with indigenous languages extend beyond the makeup of the language itself to include the thought processes imbedded in the language, as well as how, when, where and for what purposes the language is used. Only then can we begin to understand what happens to an indigenous knowledge system when the language associated with that system of thought is usurped by another.[14]

This indicates that for indigenous knowledge systems to be meaningful, they must, as much as possible, retain much of their "paraphernalia" of language, environment and mindset.

Scholars in Africa are also taking the challenge seriously of identifying different indigenous knowledge systems in various parts of the continent and how they are contributing to development and sustenance in many aspects of life. In South Africa, the *INDILINGA*, an African Journal of Indigenous

Knowledge Systems, is devoted to such research. Research has covered issues such as indigenous knowledge and communal conflict resolution, women and forest development, indigenous healthcare practice, and many others.[15] Each of these research efforts was directed at finding ways in which indigenous knowledge could be harnessed to solve some problems of development.
Archbishop Desmond Tutu, the former leader of the Truth and Reconciliation Commission in South Africa, when asked where justice was in the Truth and Reconciliation Committee's proceedings, replied that there were two different kinds of justice: the retributive justice that is western and the restorative justice that is more of African understanding. He said:

> I contend that there is another kind of justice, restorative justice which was characteristic of traditional African jurisprudence. Here, the central concern is not retribution or punishment but, in the spirit of Ubuntu, the healing of breaches, the redressing of imbalances, the restoration of broken relationships. This kind of justice seeks to rehabilitate both the victim and the perpetrator, who should be given the opportunity to be reintegrated into the community he or she has injured by his or her offence.[16]

The African restorative justice that was mainly applied in the Truth and Reconciliation Commission helped prevent South Africa from degenerating into chaos after the proceedings that would have turned back the hand of development clock of South Africa. It was a clear application of indigenous knowledge for peace and progress. This commission has become a global model today in conflict resolution.

So the point is clear that indigenous knowledge exists everywhere, and especially in developing countries, that can be channeled into the development process. However, the main issue is not its existence but the utilization of its existence for the good of people. The question, therefore, is why has this opportunity not been seriously explored? Communication appears to be part of the problem but we discuss some of those issues in the subsequent sections.

Communication Strategies for Promoting IKS

There is no dearth of indigenous knowledge in developing countries. The problem is that most of the indigenous knowledge systems of peoples in developing countries are not fully identified and vigorously promoted. One possible reason for this is that media practitioners and communication researchers do not think it is a priority, especially for economic reasons. In these days of the *commodification* of news and information, media organizations would first want to ask the question, "What economic benefit exists for us in doing this?" This is a valid question. Demand and supply principles may guide decisions here. But the validity of this question would probably apply, for the most part, in the short term. Definitely, there may not be any immediate economic reward if the media took up this role. But in the long run, the story would be different

Although development communication is believed to have come as an outgrowth of agricultural extension, it grew to encompass a process of translating and communicating new knowledge in various fields to the disadvantaged population, for the improvement of their quality of life.[17] Research in development communication has covered areas from rural development to political socialization and political participation. According to Cohen, the mass media, in both their news and entertainment functions, clearly have the potential for acting as salient agents in all manner of the development process.[18] Other researchers and scholars have argued that the process by which people, including those in Africa develop cognition, attitudes and behavioral patterns, stem, to a great extent, from the media.[19] Thus, given the tried and tested roles of the mass media in various areas of human growth and development, it would follow that the media can be effective in establishing the kind of environment in which indigenous knowledge systems can be given the prominence they deserve and placed on both local and global agenda.

Media and communications campaign strategy for the identification and promotion of indigenous knowledge systems as a strong factor in development needs to be two-pronged. There must be strategy designed with the local community as target and another with the global community as target.

Communication Strategies for the Local Community

There are two fronts of engagement as media and communication strategists focus on the local community that is the custodian of the indigenous knowledge system which researchers must have identified as applicable to development. First, there needs to be a strategy for identifying the existence of these systems wherever they reside. In identifying the knowledge systems, the people to whom the systems are indigenous will, of necessity, be identified too. This is important so that the relevant cultural context in which the systems exist will be acknowledged. Thus, in a simultaneous way, the system and the people get *confirmed*. Just as in human communication where *confirmation* and *disconfirmation* refer to the extent to which you acknowledge another person confirming a structure or a people is recognizing and inferring respect for their existence.[20] For a long time, the mass media as well as communication experts have essentially *disconfirmed* indigenous knowledge systems and the people who would claim ownership of the systems. At an initial stage, indigenous knowledge systems are more important, not for the impact they have on society, but for the fact that they derive from society and therefore intrinsically social in nature. Then they need to go beyond that stage and be idealized for what they are. But confirmation is a necessary first step. This will attract more attention to the knowledge systems being confirmed and expose them to further research to further promote their usefulness and efficacy.

In fact, the efficacy of many of these knowledge systems have already been established, but remain as research reports locked away in some distant vault. As a next step, communication strategists will begin to find ways of helping the people develop a perception of control over a very important aspect of their lives and that it matters. Very few things are as reassuring and as ego-boosting as this perception of control. This disposition will help create the conducive environment in which more can be learned of any indigenous knowledge systems. Any group, community or nation state would have a higher sense of self if they are made to perceive a sense of control over their lives. And in addition, they must be made to understand that they are making some important contribution to their own development.

The communication strategies will therefore involve those modes of communication that have been used in the past in the diffusion process, in political socialization, in the enhancement of participatory development and in the elimination of ignorance. They would also include what Frank Ugboajah called *oramedia* that consist of religious groups, town meetings, age group associations and similar groups have a direct, unmediated way of disseminating information and bringing about change.[21] The problem is, therefore, not a lack of communication modes and channels, but instead, a well-articulated subject matter such as IKS. Many communication researchers may not appreciate or understand IKS, and so, need to be educated on them. With a complete understanding of IKS, communication researchers can then have materials to work with.

This notion of promoting the essence of IKS may not gain traction if everything is left at the basic level of the community. There must be some political will that can only be derived from political policy decision makers. Unfortunately, experience has shown in many countries that the parliament as exists in many developing countries, especially in Africa, would want to be seen as patterned after western democratic systems and so may not want to be identified with what they would probably consider traditional and retrogressive ideas. It may, therefore, have to take the model of policy decision making as espoused by Vilakazi, who suggested the introduction and recognition of African traditional authority ruling in conjunction with modern democracy so that issues of grave concern to the rural communities can be brought onto the agenda of state discourse.[22]

Beyond the media, another strategy to give confirmation to indigenous knowledge systems is to include them in educational curricula. Members of a society must start, at an early stage, to appreciate what they have as their knowledge systems. This way, they will not only better understand them, but also begin to develop pride in them. This will create the support base that these systems need for national and international propagation.

Communication Strategies for the Global Community

Any action here will have to begin from the premise that all ideas are local and it is from that starting point that they eventually go global. This is important as it helps explain innovations introduced from one part of the world to another especially from the global north to the global south started as indigenous knowledge systems of the source countries. They were processed and refined and then brought to developing countries as innovations meant to be adopted for development purposes. The idea of adoption of those innovations was sold to recipient nations with the aid, in part, of the media. But they started as indigenous knowledge!

Even science, as knowledge, begins with local ideas and metamorphoses into globally accepted phenomena after continuous refinement and self-correcting procedures. It was easy for western knowledge to be passed to developing countries with relative ease because the west controlled the means of communication and there was essentially a unidirectional flow of information. The global south was handicapped in their effort, assuming they made any, to transmit any ideas to the global north because they had no control over the global media of communication. And so, any "discoveries" in the south could hardly go beyond the point of origin because it suffered from the two handicaps of lack of support from the political elite, and lack of access to global media of communication. Even when the efficacy of the application of any indigenous ideas has been established by local researchers, the whole idea gets stifled for the reasons given.

Today, however, the new mode and tool of global communication, the internet, has come to the rescue. Although there is still a technological gap between the north and the south, the south is significantly more empowered, than before, to participate in the control of global information flow and drive the information superhighway.

What does this mean for designing strategies for communicating messages about indigenous knowledge to the global community for understanding, appreciation, recognition and possible adoption? It is known that internet penetration in many developing countries is far from being at par with developed countries. But most of those who can make things happen in developing countries have access to the internet. Communication strategists can design messages on IKS to meet needs of different audiences because of the establishment of sites targeting specific demographic and professional audiences. With patience and persistence, awareness will be created and the global community will look at knowledge systems with more interest. An additional approach would also be for the media to follow researchers from developing countries that present findings of their research at local and international conferences and workshops and promote such findings as newsworthy items.

Communication researchers should now begin to look at specific strategies to promote these ideas. These should now be our focus as we shift to idealizing indigenous knowledge systems and the roles they should play in the development process.

Benefits and Challenges of this Approach

No new idea or shift in ideas comes easy. There are always benefits and challenges and it is usual to try to place them on balance and see the relative advantages that new ideas have.

There are many advantages to the shift advocated here. A very significant advantage or benefit is what Patrick Dikirr referred to as "localized relevance."[23] When we recognize indigenous knowledge systems and put them to work, individuals feel recognized and empowered. Most of what Dikirr identified as benefits of indigenizing Africa's environmental conservation would also apply here. There will be a restoration of eroding kinship solidarity networks; a more genuinely grounded sense of belonging to a place will be cultivated; the people's pride in themselves will be boosted as will be their dignity in their immobilized cultural values. It could also engender virtues of different kinds of collaboration at all kinds of levels, teamwork and cooperative problem solving mechanism; and individuals and their communities will be provided with a renewed sense of hope, audacity to dream new dreams and the confidence in not only creating their own futures but also in determining their own destiny. These are laudable benefits that will be worth the effort.

On the flip side, there are also some challenges. There is the possibility that outside forces may try to frustrate the shift for reasons that include determination to protect economic as well as cultural interests and to maintain a superiority complex that appears evident in ways in which advanced countries interact with developing countries. There could also be resistance and challenges from within from local business and political elites who have aligned themselves with the interests of their western partners and, therefore, will not provide the moral and financial wherewithal to move the effort forward. Then there is also the process of getting the local population to appreciate their new reality which may be long and drawn. That could be discouraging. Message design and choice of media will also not come easy especially as one tries to balance the traditional and the new modes of communication. However, in spite of all these, the approach is still worth trying.

Conclusion

The argument for shifting research focus to indigenous knowledge systems is not to suggest that IKS will constitute a panacea for problems of developing countries. It cannot be, because no one approach is a necessary and sufficient condition for that to happen. But it has a great potential for making a very significant contribution to the development process in many developing countries. If many communities have depended, to some degree, on products from and application of their indigenous knowledge to solve a number of problems, then the potential usefulness of these knowledge systems to help solve development problems on a larger scale should not be ignored. IKS should not be jettisoned because it is perceived as inferior. The goal now should be how to design strategies to bring IKS to the fore for what they are and what they can do. Communication scholars and researchers should summon the courage to blaze the trail.

Notes

1. Mazrui, *Cultural Engineering and Nation*, Evanston, IL: Northwestern University Press, 1972.
2. Lele, *The Design of Rural Development*, Baltimore, MD: Johns Hopkins Press, 1975.
3. Leipziger, *Basic Need and Development*, Cambridge, MA: O.G. & H Publishers, 1981.
4. Stelzer, "Question of Linkage," *The National Interest* 35 (1994): 29.
5. Sand, "Participatory Development," *Countries in Transition: Development and Cooperation* 3 (1994): 8–12.
6. Wang and Dissanayake, *Continuity and Change*. Norwood, NJ: Ablex Publishing, 1984.
7. Nandy, *Traditions, Tyranny and Utopias*. New Delhi, India: Oxford University Press, 1987.
8. Frey, Botan, Friedman, and Kreps, *Investigating Communication*. Englewood Cliffs, CA: Prentice Hall, 1991.
9. Warren, *Using Indigenous Knowledge*. Washington, DC: World Bank, 1991: 1.

10. Flavier et al., "Regional Program for the Promotion," in *The Cultural Dimension*, ed. Warren, Slikkerveer, and Brokensha, 479–487. London: Intermediate Technology Publications, 1995.

11. Lambert, "From 'Savages' to Scientists," *Tribal College Journal of American Indian Higher Education* 15 (2003): 11–12.

12. Barnhardt and Kawagley, "Indigenuos Knowledge Systems," *Anthropology and Education Quarterly* 36 (2005): 8–23.

13. Kawagley, *A Yupiaq World View*, Prospect Heights, IL: Waveland Press, 1995.

14. Smith, "Place-based Education," *Phi Delta Kappan* 84 (2002): 584–594.

15. Izugbara, Ugal, and Kinuabeye, "Indigenuous Knowledge and Communical," *INDILINGA African Journal of Indigenous Knowledge Systems* 2 (2003): 1–14.; Fasola, "Women and Forest," *INDILINGA African Journal of Indigenous Knowledge Systems* 2 (2003): 37–44; and Miah and Chowdhury, "Indigenous Healthcare," *INDILINGA African Journal of Indigenous Knowledge Systems* 2 (2003): 61–74.

16. Bell, *Understanding African Philosophy*, New York: Routledge.

17. Brace, "Coming of Age," *Media in Education and Development* 2 (1984): 78–82.

18. Cohen, *Press and Foreign Policy*. Princeton, NJ: Princeton University Press, 1963.

19. Connel, "Political Socialization," *Public Opinion Quarterly* 36 (1972): 323–333; Jennings and Niemi, *Political Character of Adolescence*. Princeton, NJ: Princeton University Press, 1974; and Chaffee, "Mass Communication and Political Socialization," in *Handbook of Political Socialization*, ed. Renchen, 223–258. NY: Free Press, 1977.

20. Devito, *Essentials of Human Communication*, Boston, MA: Pearson Custom Publishing, 2006.

21. Ugboajah, "Oramedia in Africa," in *Mass Communication, Culture, and Society in West Africa*, ed. Ugboajah, 165–176. London: Hans Zell Publishers.

22. Vilakazi, "Africa and the Problem," *INDILINGA African Journal of Indigenous Knowledge Systems* 2 (2003): 27–36.

23. Dikirr, "Challenge of Indigenizing," *Journal of Pan African Studies* 2 (2008): 81–90.

Bibliography

Barnhardt, R. and Kawagley, A. O. "Indigenous Knowledge Systems and Alaska Native Ways of Knowing," *Anthropology and Education Quarterly*, 36 (2005): 8—23.

Bell, R.H. *Understanding African Philosophy: A Cross-Cultural Approach to Classical and Contemporary Issues*. NY: Routledge, 2002.

Brace, W.E. "The Coming of Age of Development Communication."*Media in Education and Development*, 2 (1984):78–82.

Chaffee, S. "Mass Communication and Political Socialization," in *Handbook of Political Socialization*, ed. S. Renchen, 223–258. NY: Free Press, 1977.

Cohen, B. *The Press and Foreign Policy*. Princeton, NJ: Princeton University Press, 1963.

Connel, R.W. "Political Socialization and the American Family: The Evidence Re-examined" *Public Opinion Quarterly*, 36 (1972): 323–333.

Devito, J. A. *Essentials of Human Communication*. Boston, MA: Pearson Custom Publishing, 2006.

Dikirr, P. M. "The Challenge of Indigenizing Africa's Environmental Conservation Goals," *Journal of Pan African Studies* 2 (2008): 81–99.

Fasola, T. A. "Women and Forest Development in Nigeria," *INDILINGA African Journal of Indigenous Knowledge Systems* 2 (2003): 37–44.

Flavier, J.M., Jesus, A., and Navarro, C. "The Regional Program for the Promotion of Indigenous Knowledge in Asia," in *The Cultural Dimension of Development: Indigenous Knowledge Systems*, ed. D. Warren and L. Slikkerveer, and D. Brokensha, 479–487. London: Intermediate Technology Publications, 1995.

Frey, L.R., Botan, C.H., Friedman, P.G., and Kreps G.L. *Investigating Communication: An Introduction to Research Methods*. Englewood Cliffs: Prentice Hall, 1991.

Izugbara, C. O., Ugal, G.A., and Kinuabeye, J. "Indigenous Knowledge and Communal Conflict Resolution: Evidence from Nigeria," *INDILINGA African Journal of Indigenous Knowledge Systems* 2 (2003): 1–14.

Jennings, M.K. and R.G. Niemi, '*The Political Character of Adolescence*. Princeton, NJ: Princeton University Press, 1974.

Kawagley, A.O. *A Yupiaq World View: A Pathway to Ecology and Spirit*. Prospect Heights, IL: Waveland Press, 1995.

Lambert, L. "From 'Savages' to Scientists: Mainstream Science Moves Towards Recognizing Traditional Knowledge." *Tribal College Journal of American Indian Higher Education* 15 (2003): 11–12.

Leipziger, D.M. *Basic Need and Development*. Cambridge, MA: O.G.& H Publishers, 1981.

Lele, U. *The Design of Rural Development: Lessons from Africa*. Baltimore: Johns Hopkins University Press, 1975.

Mazrui, A., *Cultural Engineering and Nation Building in East Africa*. Evanston, IL: Northwestern University Press, 1972.

Miah, M.D. and Chowdhury, M., "Indigenous Healthcare Practice Through Medicinal Plants from Forests by the Mro tribe in Bandarban region, Bangladesh." *INDILINGA African Journal of Indigenous Knowledge Systems* 2 (2003): 61–74.

Nandy, A. "Traditions, Tyranny and Utopias." *Essays in the Politics of Awareness*. New Delhi, India: Oxford University Press, 1987.

Sand, van der, K. "Participatory Development and Good Governance in Countries in Transition." *Development and Cooperation* 3 (1994): 8–12.

Smith, G. "Place-based Education." *Phi Delta Kappan* 84 (2002): 584–594.

Stelzer, I. "A Question of Linkage, Capitalism, Prosperity and Democracy." *The National Interest* 35 (1994): 29.

Ugboajah, F. "Oramedia in Africa," in *Mass Communication, Culture and Society in West Africa*, ed. F. Ugboajah, 165–176. London: Hans Zell Publishers, 1985.

Vilakazi, , H.W. "Africa and the Problem of the State: Can African Traditional Authority and the Western Liberal State be Reconciled?" INDILINGA *African Journal of Indigenous Knowledge Systems* 2 (2003): 27–36.

Wang, G. and Dissanayake, W. *Continuity and Change in Communications Systems: An Asian Perspective.* Norwood, NJ: Ablex Publishing Corp., 1984.

Warren, D.M. *Using Indigenous Knowledge in Agricultural Development* (World Bank Discussion paper No. 127). Washington, DC: The World Bank, 1991.

Part 2

Theorizing and Expanding the Realm

Chapter 5

Another View: Theorizing Development Media

Chuka Onwumechili and Ritchard M'Bayo

When Denis McQuail in 1987 proposed a Development Media theory, he acknowledged difficulty by stating, "It is not easy to give a short, general statement of an emerging body of opinion and prescription appropriate to the media situation of developing countries, because of the great variety of economic and political conditions and the changing nature of situations."[1] He made the comment knowing that he was theorizing, at the time, about an emerging view of media systems affected by changing political and economic conditions in African countries. Nevertheless, McQuail's ideas have gained currency overtime, unchanged, and dominating not only debate but research on development media in Africa. McQuail was advocating new media system theories as he argued that the initial four theories of the press provided by Siebert, Peterson, and Schramm as far back as 1956, were no longer adequate and could not fully explain media systems theories that emerged all over the world. Siebert, Peterson, and Schramm had simplified understanding of how media systems operate in different parts of the world by creating four categories defining such operations. These categories were authoritarian, Soviet-communist, libertarian-free, and social responsibility which emerged from the political environment existing at the time. The authoritarian theory explained a censored press under a dictatorship, the Soviet-Communist theory explained a press under control of a Communist party, the Libertarian-Free Press theory governed a free press allowed to report as it pleases, and the Social Responsibility theory acknowledged the need to balance a free press and the need for such media to serve the public interest. McQuail argued that none of the four categories adequately explained media systems under newly independent states seeking quick economic progress. But McQuail was not alone as criticisms have come from other scholars including Altschull and Yin.[2] In each case, the critic reached similar conclusions as McQuail that the four theories put forward in 1956 were no longer adequate in explaining various media systems existing in the world. Of course, this claim was not entirely unexpected considering that Siebert, Peterson, and Schramm's seminal work was concluded several years before political independence was granted to several states in Africa and Asia. In essence, both the passage of time and the dynamic media environment were enough to change the media systems several years after 1956.

It is this passage of time and changes in the development communication environment that lead to our re-evaluation of development media theory advanced by McQuail in 1987. After all, it is more than three decades since McQuail's theory first appeared and, moreover, McQuail acknowledged that changing political and economic conditions impact theories. Our initial criticism of McQuail's work appears in a chapter in a 2011 book titled *Political Culture, Cultural Universals, and the Crisis of Identity in Africa*. Here, we take another look at that 2011 criticism in order to further reiterate our theoretical propositions and make a few modifications. In order to achieve that goal, this chapter reviews McQuail's Development Media theory in the light of reality of development media and communication in Africa. Additionally, we point to significant social changes that impact the propositions that McQuail advanced in 1987. These reviews ultimately lead us to new theoretical propositions as well as a model that provides us with an illustrative view of what we advance as the *Chumba* development media theory (CDMT). The name *Chumba* is derived from the first three letters of *Chuka* and *M'Bayo*, respectively, the names of the authors.

Reviewing McQuail's Development Media Theory

Though McQuail is often credited with the initial theory for development media, there are other scholars who had theorized about the concept before McQuail. One of them was Altschull who argued that there were fundamentally three media systems instead of the four initially advanced by Siebert, Peterson, and Schramm.[3] Altschull's three systems were

Market, Marxist, and Advancing. While the first two represented the systems in the West and the then Soviet Bloc respectively, the third system "Advancing" was for the developing world. The Advancing system became the precursor to the development media that was later put forward by McQuail. Altschull argued that the key goal of the Advancing system was unifying the nation. This he argued was justified by its proponents because most of the developing world that abided by the system are newer nations with ethnic fissures and intent on building new nation states from disparate goals of "ethnic nations" put together by colonial rulers.

McQuail's development media theory is remarkably similar to Altschull's Advancing system and has become the more cited of the two. McQuail acknowledges that the theory came to the fore following the great New World Information and Communication Order (NWICO) debate that emerged in the late 1970s and early 1980s at the UNESCO General Assembly.[4] In the debate, most of the developing nations complained about the imbalance of the flow of information with most information going from the North (rich nations) to the South (poor nations) and, thus, sought to reduce the flow by controlling the media.[5] McQuail argues that most of the South had several development needs in common pertaining to mass media use. It was his conceptualization of those common needs that led to his proposal of the development media theory. Below we list key principles that serve as the frame for the theory:

1. Media should accept and work within national development policies.
2. That the media had responsibility to promote development, national unity, language, and culture.
3. Media should give priority to national news and news of other developing countries that share similar culture and close geographically.
4. Freedom of the media may be restricted based on economic priorities and development needs of the country.
5. That developing countries considered national development as critical and justified the restriction or control of the media to ensure focus on national development priorities.[6]

Thomas McPhail concluded that McQuail's development media theory ". . .encourages an engineered press—a press committed to government-set priorities and objectives. It assumes that all efforts, including local media, need to work in unison to support national goals."[7] Indeed, this view of development media was prevalent in several countries and was hardly differentiated in countries with different governing ideologies. For instance, the socialist African countries controlled national media and in non-socialist countries the media, including privately-owned ones, were also expected to support the national development program of the government. It was clear, therefore, that this conception of a development media theory indeed reflected the interpretation of media role in several African countries immediately after political independence. For instance, the Ghanaian President at the time, Kwame Nkrumah, as well as a leading Kenya newspaper editor, Hilary Ng'weno, shared similar interpretations.[8] In fact, Ng'weno argued that press freedom was not justified in a country that had development priorities and the press in such a country had no choice but to fully support national unity and development. Moreover, Oling-Otoo and Jacobson have argued that tenets of the theory were indeed part of the policies governing the continent-wide Pan African News Agency (PANA).[9] These policies required PANA to actively seek unity and solidarity of Africa and promote preservation of African culture.

Post-McQuail: Testing, Critiquing & Impact of Social Changes

McQuail's Development Media theory is essentially normative i.e. it proposes what ought to be the case for operation of development media.

Nkrumah, Ng'weno, and Oling-Otoo & Jacobson, all cited above, provide support for this normative perspective as well.[10] Here, we refer to each of them focusing on how the media *ought* to operate or the principles laid out for such operation. They do not focus on how the media *actually* operate. It is important to stress, however, that in certain cases there are empirical examples that support McQuail's propositions. For instance, Uche's case study of the *War Against Indiscipline* (WAI) campaign in Nigeria is a classic example of actual operation of development media based on McQuail's theory.[11] Uche notes that "the private media, mostly print media (at the time), were co-opted and hijacked to join the campaign" against indiscipline conducted by then Nigerian military dictatorship.[12] Uche concluded that total media support for the campaign was a reason for its success which led to the following outcome; "Workers of all categories suddenly changed their attitudes toward work. Lateness was replaced by punctuality; . . . Instead of chaos, rudeness, and near-riot situations or rowdiness in public places, orderliness, queuing and civilized manners became the norm."[13] However, the question is whether this development media success was always the case? In this section, we answer this question as well as provide a critique of the theory before discussing major changes in the African social environment that warrant a rethinking of development media theory.

Testing McQuail's Theory and a Critique

Unlike the WAI campaign reported by Uche in 1991, there are other studies which found that privately-owned media did not always support government's development media campaigns. Here we cite two studies, one in 1991 by M'Bayo and the other in 2010 by Udomisor and Bwala. M'Bayo studied several newspapers in Sierra Leone investigating a focus on development news. What he found was startling and he concluded, ". . . development issues generally do not receive any significant media attention in Sierra Leone and, hence, communication media in the country are not development-oriented."[14] This was in spite of the fact that the government policy focused on development issues. Udomisor and Bwala found similar results in Nigeria noting that "while the government requires complete allegiance of their media organs to their policies of information dissemination, which may likely result in 'government say so coverage,' the private media owners and their papers give undue priority to commercial matters for profit motives."[15]

The studies mentioned above, as well as several others, question validity of the development media theory. In essence, whereas the theory predicts government co-option of private media as part of development media, few studies produce support for the predicted relationship. Onwumechili and M'Bayo and Mohammed provided reasons for this disconnection between theory and study outcomes citing economic, journalism education, and historical reasons, among others.[16] Economic reasons point to lack of resources that prevent certain media organizations from participating in development media communication and instead focusing on profit making as Udomisar and Bwala found in their research. Mohammed has also cited effects of journalism education, pointing out that most journalism curriculum, at the time, did not offer a single course in development journalism but mirrored courses offered for journalism education in the West.[17] The result was several journalists graduated with the perception that journalism ideals were those of a free press and not one controlled by government or one that was bound to carry out dictates of development. This problem is somewhat related to the next point that is historical. In this case, Uche pointed to longstanding opposition role of Nigerian private newspapers that goes back to the colonial years.[18] Uche claimed, that "the Nigerian press is suspicious of government intentions in every particular respect and tries to assert its role as the fourth estate of the realm, the society's watchdog."[19] This is not unique to the Nigerian private press but is found in other African countries as well. As you recall, Uche's study actually focused on a campaign that demonstrated efficacy of the development media theory, but this historical point is an acknowledgment that there are other development media campaigns that failed in Nigeria.

Critiquing the Theory

M'Bayo and Onwumechili as well as Yin have criticized development media theory on several grounds, each of which we would elaborate.[20] Already, we mentioned the lack of support for the theory's propositions in studies testing the theory. Those studies have changed the working environment for development media and have called to question several of the theory's normative propositions. In any case, we discuss some additional critiques of the theory in this section focusing on the following points: (1) the theory's assumption that development is driven only by national government; (2) the theory ignores importance of culture; and (3) social changes are unaccounted for by the current theory.

The theory assumes that only the national government drives the development imperative. This might be true of the period immediately after political independence when African countries publicly announced periodic plans for development and, thus, there was a wide awareness of those plans. But even then, there were development plans at the rural levels that had very little to do with the government. This level of development involved local councils, unions/associations, age grade groups, among several others. In recent times, there are increasingly large numbers of non-governmental organizations (NGOs) participating in development activities that may be at rural, urban, and/or national levels. Chege noted that in 1999, the United States announced that it would channel most of its aids totaling around $700 million at the time, to Africa via NGOs rather than national governments.[21] These NGOs have strong presence in several African countries, with about 2,500 such organizations in Nigeria alone and actively involved in a variety of development issues such as health, poverty and sanitation. In Uganda, NGOs were critical to the decline of HIV/AIDS incidences. Manuh identified local associations involved in different development activities in their communities including women groups such as Corn Mill societies in Cameroon, General Union of Cooperatives in Mozambique, and several rotating credit associations in Tanzania, Gambia, and Zimbabwe.[22] Unfortunately, the development media theory put forward by McQuail does not address different sources of development as we have identified above. The theory merely assumes that development is implemented solely by national governments. It is important to address these variations in development sources as the sources may well use different types of development media for communicating messages. For instance, local town unions, age groups, or community organizations do not often have resources to employ modern mass media and instead have to rely on indigenous media or *Oramedia*[23] and perhaps leaflets and such inexpensive media in order to deliver messages.

M'Bayo and Onwumechili, Soola, and Martinussen also argued that culture is central to effectiveness of development media messages.[24] Martinussen clearly made an insightful statement noting that development discourse rarely "emphasize the role of people, the role of the individual citizens, men and women, as independent actors."[25] Martinussen's concern was that the absence of this role discussion exposed underneath power struggles. However, we note that the absence also points to dismissal of culture's centrality because focus on individual actors and their world, lives, and behaviors would necessarily make culture imperative. Awa also complained, in general, about the lack of discussion of indigenous media in development when he stated, "[Indigenous media] has not received proper attention in the literature on rural development."[26] This dismissal is evident in elements proposed in development media theory. It is that absence of culture that first led some scholars to fight back by noting the importance of traditional or indigenous media which embodies the culture of the people for whom a large part of the development messages are meant. Awa, Ugboajah, and Brokensha, Warren, and Werner have all written about the important role that indigenous media and culture play in development.[27] Here, we refer to indigenous media as including local songs, dance, drama, puppet show, storytelling, market meetings, among others as part of indigenous media and in most cases these media are used outside the control of national government and are mostly unaccounted for in McQuail's development media.

Additionally, there are important social changes in Africa since the early days of development media theory and those changes warrant a rethink of the theory. Primarily, these changes have occurred at the governing system level and in communication technologies that enhanced abilability of individuals to reach a mass of people without using traditional mass media. These changes make obsolete propositions of the development media theory, the same way social changes made obsolete the ability of the initial four theories in describing media systems in newly independent African states in the early 1960s. Many African states, after independence, had one party government with the national leader at independence becoming lifelong leader of the ruling party and country. In some states, such leaders were overthrown by the military that tightened the "noose" of governance via dictatorship. This environment, under either a one party or military dictatorship, weakened people's voices in their own governance and was the background that informed McQuail's Development Media theory. The theory assumed a strong-armed government control of development planning and dissemination of development messages. However, most African nations have become democratic since the 1990s, and, thus, national governments no longer have the strong-arm control that they had in early years after independence. This new environment makes it difficult to have a unified or centralized control of development media messages. Additionally, there are significant technological changes in the last two to three decades that make it difficult to have such unified or centralized control. Widespread access to cellphones, emergence of Internet and its web resources such as social networking capabilities as well as increased transnational media all advance distributed messaging rather than unified or centralized messaging proposed by development media theory.

Advancing a New Theory

Above, we provided a critique of McQuail's development media theory which forms the basis for the *Chumba* development media theory (CDMT) that we propose in this section. The initial theory proposed by M'Bayo and Onwumechili in 2011 has nine theorems which we discuss here in addition to three theorems that modify that initial theory.[28]

To summarize, the CDMT argues that different sources of development messages work under the assumption that concerned constituencies need development and they, essentially, pursue the same or similar goals of development. However, pursuance of similar goals does not assume use of similar media nor does it assume centralized control of these media. However, an official government opposition media, unlike independent media, is likely to oppose government messages. Importantly, factors such as proximity and ethnicity are key elements that determine what the media would focus on, why and how such a focus would take place. Figure 5.1 provides a graphical representation of the theory.

The relationship shown in Figure 5.1 indicates existence of multiple sources for development messages. The messages focus on issues like health, education, among others. These messages are, ultimately, sent as announcements or information to intended receivers located in a community, or at rural, regional and/or national levels using a variety of mass media (and/or interpersonal) modes that are controlled by government, private organizations, and/or individuals. Activities at the receiver level, that may demonstrate impact of development messages, are fed back to initial sources of development messages and to other receivers as media reports or testimonials. The feedback may also go directly from receivers to funding sources as organizational reports.

Figure 5.1: Model of *Chumba* Development Media Theory (CDMT)

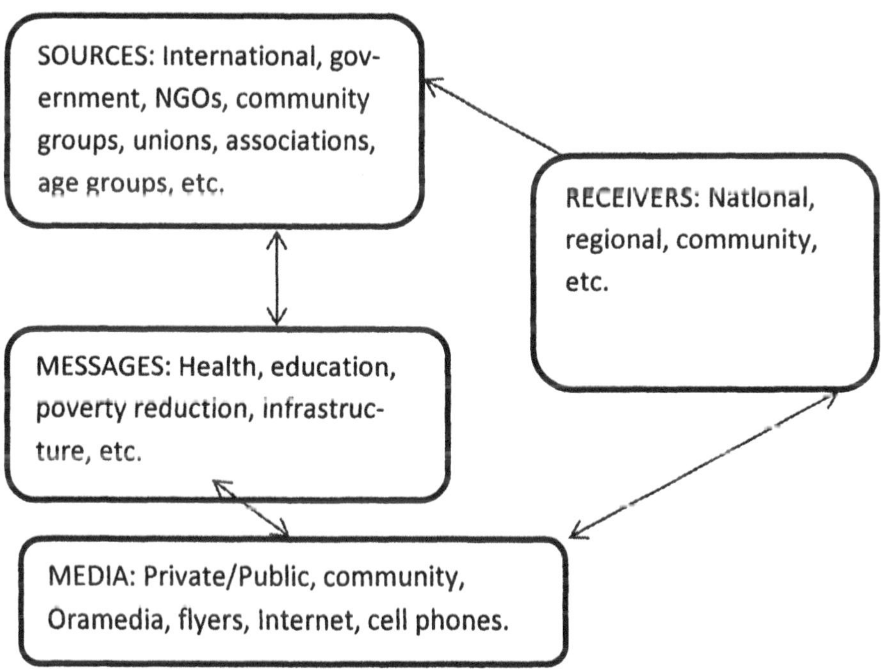

It is based on the above understanding that we advanced nine theorems of the CDMT which first appeared in M'Bayo and Onwumechili. Below we list each theorem with a brief empirical example for support.

> T1: African mass media perspectives on development issues and their support for the same are related to government ownership or leverage over mass media.

Odhiambo demonstrated that Kenyan government ownership of a network of rural newspapers was critical in government development policies receiving support from rural newspapers whose workers served at the pleasure of the government.[29] Several other studies also confirm this link including studies by Udeze and Onwumechili and M'Bayo.[30]

> T2: Independent media in an African state, under one-party or military rule will, in the main, share government perspectives on development issues.

Here, the reference to "independent" is fleeting and it is more accurate to simply label these media organizations as non-government owned. Uche's work demonstrates how the private media played a critical role in the success of government programs.[31] Uche noted that support from private media came because of consequences of non-support. These consequences included revocation of import licenses for newsprint. Thus, private media decision to share government perspectives may often be based on survival instincts in a harsh environment.

> T3: Opposition media or opposition-support media in a democratic multiparty African state will be critical of government on development issues.

Several examples of such media exist and they range from opposition party media to clandestine media and they have in common the goal of subverting activities, including those focused on development, of the sitting government. Thus, such media exist to critique government programming. Examples can be found in the work of Kintz and Campbell.[32]

> T4: Mass media coverage of development issues in Africa is dependent on urban proximity and relatively easy access to news and not a conscious effort to promote the primacy of culture in national development.

Okigbo, Okigbo, and Nwokeafor noted that content of African media "reflects their 'township' orientation" which results from their training, lack of resources, and easy access to urban news.[33] There are other studies cited in M'Bayo and Onwumechili that support this theorem.[34]

> T5: The mass media unintentionally chronicle development progress of a nation through its long term gradual coverage of scheduled events that are both private and public.

As argued in M'Bayo and Onwumechili, the gradual coverage of a development issue may indicate attenuation or an increase in the said issue. For instance, Nigerian media focused attention on covering individuals who had earned a bachelor's degree in the 1930s until 1950s but today this coverage no longer exists as earning such degree is no longer unique.

> T6: Mass media language is dependent on the perception of the audiences' language-of-doing-business and not on language as a vehicle of self-development and self-determination.

Though McQuail's Development Media theory had argued about the importance of promoting a national language, this is the most difficult to accomplish. Instead, development media language is based on the most spoken language which is often English, French, or Portuguese and not the indigenous languages that are not only several but also spoken by smaller fractions of the national population. However, this does not deny that indigenous language is used when development message is narrowly focused on a particular ethnic community.

> T7: Regional or ethnic-based media are more likely to cover development issues of their regions than they are to cover national development issues.

Media coverage of events, whether focused on development or not, may tilt towards ethnicism when such media seek to protect regional or ethnic interests. A classic example comes from the Rwandan genocide in 1994 when the mass media—Radio Television Libre des Mille Collines (RTLM) and the *Kangura* newspaper—were used to fan hatred among ethnic groups.[35] These types of particularism also occur with media coverage of development in several African states.

> T8: The mass media help in establishing or focusing of national development policies by setting an agenda for both government and the public.

The work of Ghana's *Statesman* newspaper in 2006 is important here. The *Statesman* went against public opinion to argue for the right of gays, lesbians, bisexual, and transgendered.[36] This agenda setting move brought the issue to the top of national agenda even though the country's Commission on Human Rights and Administrative Justice (CHRAJ) eventually refused to advocate for such rights. M'Bayo and Onwumechili also cite media efforts in bringing the issue of illegal waste dumping in the country to the attention of the government that led to public policy on health and sanitation.

> T9: The substantiation of new social classes of uneducated rich and educated poor is more likely to support potential types of new development media in Africa.

As M'Bayo and Onwumechili argued, there is an increase in the population of educated poor and uneducated rich classes in Africa. They argue that new types of centers, such as communication centers, are likely places to service development communication of both classes. The uneducated rich, particularly, have resources to purchase such services for their educational needs and development sources such as the NGOs or neighborhood groups will be the providers.

Additional Theorems

We add the following theorems to the ones identified above. These three new theorems lead to a better understanding of the CDMT. We also provide brief support for each of the three theorems.

> T10: Sources of development planning and implementation are varied and, hence, the media for disseminating development messages are also varied.

The United States African Development Foundation, McPhail, and Barrow and Jennings provide examples of numerous development projects under various sources.[37] While McPhail and Barrow and Jennings focus on the work of NGOs, the USADF identifies small-scale development projects that it sponsors through Africans in Africa. In 2008, alone, ADF spent $12.8 million on new small-scale projects in the continent. Importantly, these sources and the national government programs use a variety of media and developed different types of communication messages for their audiences.

> T11: The more development planners rely on emerging media—new media such as texting, the World Wide Web and the Internet—the more varied the message types would be.

Emetumah and Fab-Ukozor point to the use of short message service (SMS) in Kenya noting that Kenya Agricultural Commodity Exchange (KACE) operates a sort of franchise resource center that gathers information on market and products and distributes same through SMS to farmers for optimal output."[38] This new system also provides for more participatory communication instead of the top-down communication that exists with modern mass media.

> T12: The more development planners rely on personal contact media for disseminating messages, the more varied the message types would be.

This theorem focuses on an increase in use of new media which provides more personal contact with the receiver of development communication message. The more narrowly focused such messages become, the more likely they will be individually designed increasing the variety of message types.

Differentiating from Other Media Systems Theories

Advancing a new theory of development media raises several questions including—How does this new theory differ from the six theories of the press presented by McQuail in 1987? As we noted earlier, four of those theories—Authoritarian, Soviet-Communist Press, Libertarian-Free Press, and Social Responsibility Press—were initially proposed by Siebert, Peterson, and Schramm in 1956. McQuail, thereafter, added the Development Media theory discussed at length in this chapter as well as the Democratic-Participant Media Theory. In this section, we differentiate the CDMT from each of the six.

Both the Authoritarian and the Soviet-Communist Press theories are dramatically different from the proposed CDMT. The former two are based on centralized control of the media by the ruling elite—a dictator or monarch in the case of the authoritarian theory and the Communist or Socialist party in the case of the Soviet-Communist theory. In either case, the media are gravely censored by the elite. The CDMT does not dismiss the likelihood of media censorship as this is expected in government controlled media. However, the CDMT acknowledges that in several developing countries, not all media are under centralized control and, thus, the private and/or opposition media are free and may not be censored. Moreover, the CDMT recognizes existence of small and more diverse media systems that are neither controlled by the government or large organizations.

Attributes of the CDMT, described above, also differentiates it from the Libertarian-Free Press theory advanced by Siebert, Peterson, and Schramm. The Free Press theory assumes a media system with provisions ". . . to hold opinions freely, to express them. . .the theory seems designed to protect opinion and belief. . ."[39] This media system operates in environments where governments are not involved in media ownership, unlike the environments described by the CDMT. It is important here to reiterate that what is described in the proposed CDMT is one that combines both what is found in the Free Press Systems as well as some of the elements found in the Authoritarian/Soviet-Communist media theories identified in the previous paragraph.

The Social Responsibility Media Theory had come to the fore following criticisms that the Free Press does not serve the public's interest. Thus, this theory focuses on the belief that the goal of a free press must be balanced against need to serve the public interest particularly in the case of broadcasters that use frequency bands which are public property. This is justification for public hearings during license renewals for broadcasters. These media are all privately owned just as is the case with the Free Press theory discussed previously. Therefore, the same differences identified previously between the Free Press and the CDMT are also relevant here.

The Democratic-participant media theory, according to McQuail, is located mainly in developed liberal societies. McQuail argues that it focuses on ". . . the right to relevant information, the right to answer back, the right to use the means of communication for interaction in small-scale settings of community, interest group, sub-culture."[40] He also provided examples of such media as including community cable television, micro-media in rural settings, political posters, and underground press. In today's world, this would include blogs and community websites, among others. He adds, as an emphasis, that "Communication is too important to be left to professionals."[41] Ugande and Mojaye have argued, however, that the Democratic-participant media theory was descriptive of a normative development media in Africa.[42] This challenges McQuail's initial claim that the Democratic-Participant Media theory existed mainly in liberal societies of the West. However, one must acknowledge that Ugande and Mojaye's claims do not stray too far from McQuail's observation that aspects of the Democratic-participant theory had "some elements (also) present in development media theory, especially its emphasis on the 'basis' of society, on the value of horizontal rather than vertical (top down) communication."[43] Those elements are discussed in the subsequent paragraph.

We have spent most of this chapter comparing McQuail's Development Media theory with the CDMT. Instead of rehashing the points that we have made in previous sections of this chapter, we provide a summary here. First, CDMT unlike the Development Media theory, is a grounded rather than a normative theory. By grounded, we refer to its theorems emerging from empirical observation of how development communication operates which is different from a normative theory which focuses on prescribing how development media *ought* to operate. As we noted earlier, a major problem with McQuail's development media theory is that it depended on media of the 1980s whereas the media world and other social changes have taken place since then. Furthermore, the CDMT acknowledges multiple sources of development communication and also notes that these sources do not have a centralized control as proposed by the development media theory. Moreover, these sources use a variety of media communication channels that include modern broadcast and other modern mass media, indigenous media systems, and newer and emerging media systems that significantly enhance individual's control of his/her own messages to a mass of people or recipients.

Conclusion

This chapter has focused on the need for a new theory of development media. In essence, it recognizes inadequacy of the current theory, particularly as it affects understanding of development media theory in Africa. The chapter presents its argument by reviewing current development media theory advanced by Denis McQuail in 1987 and noted that at the time of its introduction, McQuail sought to fill gaps in understanding media systems in the world. Importantly, McQuail was aware that the UNESCO study in the late 1970s and early 1980s which culminated in the New World Information and Communication Order (NWICO) was a major marker identifying a call for recognition of new media systems. At the time, there were four such system theories—Authoritarian, Soviet-Communist, Libertarian-Free Press, and Social Responsibility—advanced in 1956 by Siebert, Peterson, and Schramm. A review and subsequent critique of McQuail's Development Media theory points to some weaknesses which included: (1) the theory's assumption that development is driven only by national government; (2) the theory ignores importance of culture; and (3) social changes are unaccounted for by the current theory.

Ultimately, the chapter puts forward a new theory—the Chumba Development Media Theory (CDMT)—which was initially advanced by the authors in a 2011 work. The initial CDMT included nine theorems grounded in empirical observations about the media system in Africa. In this chapter, the authors add to the nine theorems by advancing additional three theorems. Furthermore, they provide for the first time, a model to help understand relationships among critical aspects of the theory. Details of the empirical supports for the theorems can be found in the original work of the authors published in 2011. The closing section of the chapter focuses on differentiating the CDMT from each of the six media theories that include the original four proposed by Siebert, Peterson, and Schramm and the additional two advanced by Denis McQuail in 1987. The intent is to demonstrate that the CDMT is indeed different from existing theories and, importantly, that it exceeds prevailing theories in scope in terms of explaining and describing the environment for development media communication in Africa.

Notes

1. McQuail, *Mass Communication Theory*. Malden, MA: Blackwell, 1987.
2. Altschull, *Agents of Power*. NY: Longman, 1984 and Yin, *Beyond Four Theories*. Columbia, SC: AEJMC, 2008.
3. Altschull, *Agents of Power*.
4. McQuail, *Mass Communication Theory*.
5. MacBride, *Many Voices, One World*. Paris: UNESCO, 1980.
6. McQuail, *Mass Communication Theory*.
7. McPhail, *Global Communication*. Malden, MA: Blackwell, 2006, 40.

8. Nkrumah, *African Journalist*. Dar es Salaam: Tanzania Publishing, 1965, and Ng'weno, "Role of the Press." Paper presented at IPI Assembly in Nairobi, Kenya, June 4, 1969.

9. Oling-Otoo and Jacobson, "Pan-African News Agency." In *Communication and Transformation of Society*, ed. Onwumechili and M'Bayo, 151–164. Lanham, MD: University Press of America, 1995.

10. Nkrumah, *African Journalist*; Ng'weno, "Role of the Press," and Oling-Otoo and Jacobson, "Pan-African News Agency."

11. Uche, "Communication and Development." In Communication in Development, ed. Casmir, 132–156. Norwood, NJ: Ablex, 1991.

12. Uche, "Communication and Development," 139.

13. Uche, "Communication and Development," 144.

14. M'Bayo, "Information Policies in an African State." Phd diss., Howard University, 1991, 125.

15. Udomisor and Bwala, "Newspaper Coverage of Development." *Nigerian Journal of Communication* 8 (2010): 25–42.

16. Onwumechili and M'Bayo, "Structure and Functions of the Mass Media." In *Communication and Transformation of Society*, ed. Nwosu, Onwumechili, and M'Bayo, 53–75. Lanham, MD: University Press of America, 1995, and Mohammed, "Development Communication Training." *Africa Media Review* 4 (1990): 75–88.

17. Mohammed, "Development Communication Training."

18. Uche, "Communication and Development."

19. Uche, "Communication and Development," 139.

20. M'Bayo and Onwumechili, "Rethinking Development Media Theory." In *Political Culture, Cultural Universals*, ed. M'Bayo, 129–153. Lewiston, NY: Edwin Mellen, 2011.

21. Chege, "Donors Shift More Aid to NGOs." Africa Recovery Online 13 (1999): www.un.org/

22. Manuh, "Women in Africa's Development." Africa Recovery Online 11 (1998): www.un.org/

23. Oramedia first coined by Frank Ugboajah in 1985 refers to indigenous media. In the case of Africa, such media are mainly oral and hence the name *Ora(l)media*.

24. M'Bayo and Onwumechili, "Rethinking Development Media Theory"; Soola, "Development Communication: The Past." In *Communication for Development*, ed. Soola, 1–27. Ibadan: Kraft Books, 2002; and Martinussen, *Society, State, and Market*. New York: Zed Books, 1997.

25. Martinussen, *Society, State, and Market*, 331.

26. Awa, "Role of Indigenous Media." In *Communication and Transformation of Society*, ed. Nwosu, Onwumechili, and M'Bayo, 237–251. Lanham, MD: University Press of America, 1995, 243.

27. Awa, "Role of Indigenous Media"; Ugboajah, *Mass Communication, Culture and Society*; and Brokensha, Warren, and Werner, *Indigenous Knowledge Systems.* Lanham, MD: University Press of America, 1980.

28. M'Bayo and Onwumechili, "Rethinking Development Media Theory."

29. Odhiambo, "Development Journalism in Africa." *Africa Media Review* 5 (1991): 17–29.

30. Udeze, "Impact of Government Ownership." In *Contemporary Issues in Communication*, ed. Ndolo, 219–229. Enugu, Nigeria: Rhyce Kerex, 2011 and Onwumechili and M'Bayo, "Structure and Functions of the Mass Media."

31. Uche, "Communication and Development."

32. Kintz, "Overview of Current Media." *Global Media Journal* 6 (2007) and Campbell, *Emergent Independent Press*. Westport, CT: Praeger, 1998.

33. Okigbo, Okigbo, and Nwokeafor, "Strategic Communication and Education." *Journal of African Communications* 4 (2002): 1–31.

34. M'Bayo and Onwumechili, "Rethinking Development Media Theory."

35. Mgbako, "Rwanda: Media Censorship." (2010): allafrica.com/

36. AllAfrica.com, "Publisher Sounds Alarm." (2006). www.allafrica.com/

37. USADF, "Country Programs." www.adf.gov/; McPhail, *Global Communication;* Barrow and Jennings, *Charitable Impulse*. Oxford: James Currey, 2001.

38. Emetumah and Fab-Ukozor, "ICT's Effectiveness." In *Perspectives on Communication and Culture*, ed. Wilson, 133–143. Uyo, Nigeria: BSM Resources, 2010, 142.

39. McQuail, *Mass Communication Theory*, 113, 115.

40. Ibid, 122.

41. Ibid, 123.

42. Ugande, "Community Radio." In *Media, Governance and Development*, ed. Oso, Ajibola, and Odunlami, 558–568. Ago-Iwoye: Idowu Sobowale Conference, 2009 and Mojaye, "Needs-based Communication." In *Globalization and Development*, ed. Mojaye, Oyewo, M'Bayo, and Sobowale, 55–64. Nigeria: Ibadan University Press, 2008.

43. McQuail, *Mass Communication Theory*, 122.

Bibliography

AllAfrica.com. "Publisher Sounds Alarm and Points to Looming Titanic Clash." (2006, Sept. 28): allafrica.com/stories/200609280563.html.

Altschull, J. *Agents of Power: The Role of the News Media in Human Affairs*. NY: Longman, 1984.

Awa, N. "The Role of Indigenous Media in African Social Development." In *Communication and the Transformation of Society: A Developing Region's Perspectives,* ed. Nwosu, P., Onwumechili, C., and M'Bayo, R., 237–251. Lanham, MD: University Press of America, 1995.

Barrow, O., and Jennings, M. *The Charitable Impulse: NGOs and Development in East and North East Africa.* Oxford: James Currey, 2001.

Brokensha, D., Warren, D., and Werner, O. *Indigenous Knowledge Systems and Development.* Lanham, MD: University Press of America, 1980.

Bryant, C., and White, L. *Managing Development in the Third World.* Boulder, CO: Westview Press, 1982.

Campbell, J. *The Emergent Independent Press in Benin and Cote d'Ivoire: From Voice of the State to Advocate of Democracy.* Westport, CT: Praeger Publishers, 1998.

Chege, S. "Donors Shift More Aid to NGOs: But Cooperation Between Governments and NGOs is Critical for Greater Effectiveness." Africa Recovery Online (A UN Publication), 13(1999): www.un.org/ ecosocdev/ geninfo/ afrec/ vol 13 no1/jun99.htm

Ciboh, R. "Newspaper Coverage of Health Issue in Nigeria." *The Nigerian Journal of Communication* 8(2010): 220–241.

Emetumah, F., and Fab-Ukozor, N. "ICTs Effectiveness in Rural Development in Kenya." In *Perspectives on Communication and Culture*, ed. Wilson, D., 133–143. Uyo, Nigeria: BSM Resources Nigeria Ltd, 2010.

Kintz, L. "Overview of Current Media Practices and Trends in West Africa." *Global Media Journal, 6* (2007):

MacBride, S. *Many Voices, One World: Communication and Society Today and Tomorrow.* Report by the International Commission for the Study of Communication. Paris: UNESCO, 1980.

Manuh, T. "Women in Africa's Development: Overcoming Obstacles, Pushing for Progress." Africa Recovery (A UN Publication), 11 (1998):.www. un. org / ecosocdev/geninfo/afrec/.

Martinussen, J. *Society, State, and Market: A Guide to Competing Theories of Development.* NY: Zed Books, 1997.

M'Bayo, R. "Information Policies in an African State: A Study of the Rules and Regulations of Communication in Sierra Leone." Doctoral dissertation, Howard University, Washington, D.C, 1991.

M'Bayo, R., and Onwumechili, C. "Rethinking Development Media Theory Through Normative Propositions and Social Scientific Theorems." In *Political Culture,Cultural Universals, and the Crisis of Identity in Africa: Essays in Ethnoglobalization*, M'Bayo, R.,129–153. Lewiston, NY: Edwin Mellen Press, 2011.

McPhail, T. *Global Communication: Theories, Stakeholders, and Trends.* Malden, MA: Blackwell Publishing, 2006.

McQuail, D. *Mass Communication Theory: An Introduction.* Beverly Hills, CA: Sage Publications Inc, 1987.

Mgbako, C. "Rwanda: Media Censorship will Breed Resentment." (2010, Aug. 11): allafrica.com/stories/201008110001.html.

Mohammed, J. "Development Communication Training in Nigeria: Notes and Observations." *Africa Media Review*, 4(1990), 75–88.

Mojaye, E. "Needs–based Communication for People–Centered Development of Nigeria's Niger Delta. In *Globalization and Development Communication in Africa,* ed. Mojaye, E., Oyewo, O., M'Bayo, R., and Sobowale, I., 55–64. Nigeria: Ibadan University Press, 2008.

Ng'weno, H. "The Role of the Press in a Developing Country." Paper presented at the IPI Assembly in Nairobi, Kenya, June 4, 1969.

Nkrumah, K. *The African Journalist.* Dar es Salaam: Tanzania Publishing House, 1965.

Odhiambo, L. "Development Journalism in Africa: Capitulation of the Fourth Estate?" *Africa Media Review*, 5(1991): 17–29.

Okigbo, C.C., Okigbo, C. A., and Nwokeafor, C. "Strategic Communication and Education for Community Development in Africa." *The Journal of African Communications*, 4 (2002): 1–31.

Oling-Otoo, M., and Jacobson, R. "The Pan-African News Agency: A Manifestation of Development Media Theory." In *Communication and the Transformation of Society: A Developing Region's Perspectives*, ed. Nwosu, P., Onwumechili, C., and M'Bayo, R., 151–164. Lanham, MD: University Press of America, 1995.

Onwumechili, C., and M'Bayo, R. "Structure and Functions of the Mass Media in National Development in Africa: Some Systemic Considerations." In *Communication and the Transformation of Society: A Developing Region's Perspectives,* ed. Nwosu, P., Onwumechili, C., and M'Bayo, R., 53–75. Lanham, MD: University Press of America, 1995.

Servaes, J. "Toward a New Perspective for Communication Development." In *Communication in Development*, ed. Casmir, F., 51–85. Norwood, NJ: Ablex Publishing Corporation, 1991.

Siebert, F., Peterson, T., and Schramm, W. *Four Theories of the Press*. Urbana, IL: University of Illinois Press, 1956.

Soola, E. "Development Communication: The Past, the Present, and the Future." In *Communication for Development,* ed. Soola, E., 1–27. Ibadan: Kraft Books Ltd, 2002.

Uche, L. "Communication and Development in a Military Political Culture: A Case Study of a Nigerian Public Campaign for Social Change." In *Communication in Development*, ed. Casmir, F., 132–156. Norwood, NJ: Ablex Publishing Corporation, 1991.

Udeze, S. "The Impact of Government Ownership on Print Medium Management: A Study of Daily Star Enugu." In *Contemporary Issues in Communication and Society*, ed. Ndolo, I., 219–229. Enugu, Nigeria: Rhyce Kerex Publishers, 2011.

Udomisor, I., and Bwala, S. "Newspaper Coverage of Development and Cultural Issues in Nigeria." *The Nigerian Journal of Communication* 8(2010): 25–42.

Ugande, G. “Community Radio, Participatory Broadcasting and Popular Empowerment.” In *Media, Governance, and Development in Africa: A Book of Proceedings*, ed. Oso, L., Ajibola, T., and Odunlasmi, D., 558–568. Ago-Iwoye: Idowu Sobowale Conference on Media Governance and Development in Africa, 2009.

Ugboajah, F. *Mass Communication, Culture and Society in West Africa*. Oxford: Hans Zell, 1985.

United States African Development Foundation. (Undated). USADF Country programs. (Undated): www. adf.gov/documents/ADF_Bochurefinal 109. pdf.

Yin, J. *Beyond the Four Theories of the Press: A New Model for the Asian and World Press*. Columbia, SC: AEJMC, 2008.

Chapter 6

The Theory of Communicative Action for Population-Communication Management in Sub-Saharan Africa: Argumentative Speech in Nigeria

Cornelius B. Pratt and Thomas L. Jacobson

This chapter applies Habermas's theory of communicative action to three demonstrations of a fertility-reduction program to facilitate participants' knowledge of fertility issues presented in a discourse that serves as a process toward reaching a consensus-bringing force of argumentative speech in Nigeria. Such an analysis is justified by moderate declines in population growth and fertility rates in several African countries, even as their governments still describe those rates as "too high." Population-communication management specialists, policy and campaign actors and their targets can negotiate three universal, criticizable validity claims: propositional or constative truth; normative rightness; and subjective truthfulness, sincerity or authenticity—all of which must be well formed or comprehensible (that is, linguistically intelligible) and made under symmetry conditions of ideal, unconstrained speech. We begin with the following quotation:

> The projected [population] figures for sub-Saharan Africa are the most problematic, climbing from 10 per cent of the world's population in 2000 to 17 per cent in 2050 and to 21 per cent in 2100 before remaining constant at around one-fifth of the global population.

—United Nations.[1]

> The unmet need for family planning remains moderate to high in most regions, particularly in sub-Saharan Africa . . .
> —United Nations.[2]

Africa, particularly the sub-Saharan region, is understandably in a frenetic pace to develop its human capital, natural resources, and global presence politically, and to establish a firm toehold for participating in the global marketplace. Consequently, it has been making significant investments in revamping its educational institutions, health-care systems, infrastructure, social services, and its national security. Even so, a number of sub-Saharan African countries are at the precipice of an economic implosion and political stalemate. The region's once-massive swaths of post-colonial national developments (e.g., in the Central African Republic, the Democratic Republic of the Congo, Liberia, Nigeria, Sierra Leone, and Zimbabwe) have succumbed to dystopia, exposing societies to their share of economic and political recidivism. The Democratic Republic of the Congo, for example, is still a one-person dictatorship even after the death in 1997 of its autocratic leader, Mobutu Sese Seko, who governed the country for thirty-two years. Nigeria is now one of the world's largest importers of staple foods, of which the country was once a major exporter. Sixty-five percent of the country's population is food-insecure, which means that about 97 million Nigerians are at risk of finding out that they do not have anything to eat. Zimbabwe, under Ian Smith's government, was the bread basket and political bellwether of much of the continent. But its present government's economic policies fomented a disastrous land-reform policy, enabled the death of its currency (the Zimbabwean dollar), and fostered an 85 percent unemployment rate and spiraling hyperinflation in sextillionth proportions. Sierra Leone, long an exemplar of democratic governance on the continent, is confronted by a massive task of reinventing itself, following years of ethnic violence and political instability. Thus, "[m]uch of Africa has stagnated and even regressed in terms of foreign trade, investment, per capita income, and other economic growth measures. A few countries such as Angola, Botswana and Mozambique have somehow managed to keep their economies growing."[3]

Africa's development challenges are largely an outcome of population issues. The purpose of this chapter, therefore, is to apply the theory of communicative action (TCA) as a complement to existing public-engagement approaches to population-communication management

(PCM) in Africa. It draws upon some of the key *universal* elements of TCA to help guide the 75 percent of African governments that "viewed [their nations'] fertility rates as too high" and to enable them manage more effectively communications undertaken specifically to respond to population issues.[4] Those universal elements could also be applied to other national or community-development programs on the continent. And TCA could better enable "government institutions to integrate population dynamics into policymaking and the development of poverty-reduction plans and expenditures."[5]

Sachs acknowledged the cultural bases of high fertility rates in poor countries: age of marriage, expectations about family size, onset of childbearing age, lag between births, and the unpopularity of contraception.[6] And he presented four compelling reasons for today's poorest countries to speed up a demographic transition: (a) poor families cannot overcome poverty without a decline in fertility rate, (b) poor countries with high population growth rates cannot achieve economic progress, (c) poor subsistence farmers with high population growth rates face shrinking farmlands and falling incomes, and (d) rapid population growth rates raise mass migration and the possibilities for violence and local conflicts. He averred that a voluntary reduction in fertility rates, especially in countries least able to absorb a rising population, was crucial to stabilizing world population on an already crowded planet at eight billion or fewer by 2050, rather than allowing it to grow from its current 6.7 billion to 9.2 billion. If, however, the current total fertility rate (TFR) remains unchanged between now and 2050, the world's population will balloon to 11.7 billion. A continuing high fertility rate in sub-Saharan Africa undermines Millennium Development Goals-related projects, such as improving access to reproductive health; reducing child mortality, hunger and malnutrition; and increasing primary education enrollment.[7]

Cast against the continent's economic and national-development gains and losses are growth indices that further threaten the sustainability of the region's sometimes buoyant, sometimes teetering economies. Africa's fertility rates, that is, the total number of children the average woman in a population is likely to have throughout her life, based on current birth rates, are, for one, still generally high, except in southern Africa, which has less than one-thirteenth of the continent's population.[8] The continent has eight of the world's top-ten fertility rates.[9] The continent has twenty-one of twenty-five nations with the highest rates. While fertility has been declining in recent decades in sub-Saharan Africa, the pace of the decline is converging slowly toward the world mean total fertility rate.[10]

For another, Africa has the world's highest estimated annual population growth rate of 2.4 percent (2.5 percent for the sub-Saharan region), even as the continent has more malnutrition, conflicts and social instability and more unstable economies than any other region. Therefore, it seems logical for African governments to address looming birth and population growth rates. In 1976, 25 percent of African governments had policies aimed at lowering their fertility; by 2007, that number had increased to 72 percent.[11] Similarly, in 2007, 64 percent of African government reported having policies aimed at reducing the rapid growth of their respective populations; in 1996, that number was 60 percent, in 1986 39 percent, and in 1976 25 percent. In essence, African governments have an increasing interest in resolving their population issues.

Because African nations have population-based demographic challenges that have major implications for their economic growth and sustainable development, political stability, healthcare delivery programs, education policies, and their overall national wellbeing, it is imperative that their governments formulate campaigns that meet informational and motivational objectives on increasing public awareness of and on responding strategically to the continent's excruciating population issues. Fertility transitions are occurring in Ghana, Kenya, Senegal, the southern half of Nigeria, all of southern Africa, including Zimbabwe; and effective family planning programs have been instrumental in enabling South Africa to embark upon a fertility transition comparable to those of developing regions outside of Africa.[12] A variety of activities have helped engender such outcomes. For example, first marriage is delayed, particularly in southern Africa[13]; births are postponed and spaced[14]; and the number of never-married women is increasing throughout the continent. But population growth, as indicated in high premarital fertility, challenges existing population policy.[15]

More recently, Connelly criticized family planning as ineffective for population control, linked the latter to imperialism, and argued that it was the emancipation of women that remade humanity, not population control whose policies he described as coercive.[16] He wrote, "The great tragedy of population control, the fatal misconception, was to think that one could know other people's interests better than they knew it [*sic*] themselves."[17] Similarly, Bodunde took issue with the term "population control," suggesting "responsible parenthood" as a substitute and advocating family planning messages be made more attractive to Nigerian men, who typically dictate family size.[18] But current family planning programs are weak and their future performance is questionable, even as such programs are a cost-effective approach to reducing population pressures.[19]

The Reconstructive Theory of Communicative Action

Habermas, perhaps the most widely known contemporary German philosopher and a critical theorist of the Frankfurt School, distinguishes between genuine and false consensus through his emphasis on what he describes as communicative action, a symmetry condition in which there is unconstrained dialogue to which all speakers have access, and in which they have the prevailing force of better judgment.[20] Communicative action is defined in relation to "the general symmetry conditions that every competent speaker who believes he is engaging in an argumentation must presuppose as adequately fulfilled."[21] Those conditions are that (a) there must be a symmetrical distribution of opportunities to contribute to discussions, i.e., all can speak; (b) there must be opportunities for participants to raise any and all subjects they wish to see addressed, i.e., nothing is "off the table"; and (c) there must be an inviting environment to discuss every topic fully and to the satisfaction of those who raised the topic. Actors seek through speech acts an orientation toward understanding, consensus and agreement rather than toward merely fulfilling their own specific agendas or personal goals. Such a theoretical perspective emphasizes planning through widespread community participation, dialogue, genuine consensus, cooperation, and negotiation, not necessarily through strategic action.

Habermas focuses on interaction engaged by both senders and receivers of linguistic acts geared toward understandings that are rationally negotiated by all parties. But there are limits to such negotiations. While, say, parties to a speech act can understand the conditions under which a message recipient can understand, accept and endorse the speech act and be persuaded by its claims (e.g., "Keeping your family size manageable makes it possible for you to help your own family lead a more fulfilling life in many other ways," or "Having fewer children means having more"), neither the recipient nor the sender can know the conditions of the acceptability of the speech act relative to the actual attitudes and beliefs of the individual other.

Habermas distinguishes between strategic action, a standard tool for PCM in Africa, and communicative action, a yet to be used tool in that context. The former is goal-oriented (toward effectiveness and success); the latter, process-oriented. This dichotomy reflects what Habermas describes as the paired, contrasting concepts of the system (read: strategic action) and the life world (read: communicative action).

Parties engaged communicatively do so within the background of their life world—their taken-for-granted background convictions and assumptions and naively mastered skills, their cultural historical environment, and their repository of the interpretive work of preceding generations. To engender rational conduct of life, then, cultural tradition must allow the following conditions if attempts to preserve the life world are to be successful: (a) it must make available to parties formal concepts for the objective, social, and subjective worlds and allow differentiated validity claims and stimulate a corresponding differentiation of attitudes; (b) it must strip dogmatism, allowing cultural interpretations to be subject to questioning and critical revision; (c) it must allow feedback, supported by arguments rendered fluid through permanent criticism to the degree that such a learning process can be socially institutionalized; and (d) it must interpret the life world in such a manner that it uncouples success-oriented action from action oriented toward reaching understanding that is communicatively renewed over and over. If a society does not allow these conditions, then the life world may become subservient to the systems world, which Habermas refers to as a colonization of the life world.

In sum, six key components of TCA that readily apply to PCM are (a) strategic action; (b) communicative action; (c) rationalistic argumentation; (d) universal, criticizable validity claims; (e) the ideal speech situation; and (f) cultural life world.

The Practical Relevance of the Theory of Communicative Action—to Date

TCA has both strategic and communicative relevance to national development initiatives. First, in providing a forum to Africans for the collaborative process of crafting campaign messages generally perceived as culturally sensitive, the participation is viewed as an additional opportunity to increase the chances for the messages to be accepted politically and socially by the government and the community, respectively.[22]

Second, the interactions and exchanges can serve as a forum to change PCM goals, objectives, strategies, tactics, and protocols through presuppositions of argumentation, for which Africans have a growing penchant, and for which they have their own historical antecedents in the rhetorical traditions of Ptah-Hotep's work that predated Greco-Roman rhetoric by about 3,000 years.[23] As Blake noted, Ptah-Hotep's *The Instruction of Ptah-Hotep*, composed about 4,000 B.C., identified the earliest formulation of a rhetorical theory on "fair speech," particularly in terms of a technical skill: argumentation. Birthrates fell among Kassena-Nankana women of northern Ghana after they were empowered to engage in frank conversations with men, to exchange ideas and to participate in hitherto male-only "durbars" or political gatherings.[24] Epstein concluded: "The individual human voice, a tool available even to the poorest people, may be the key not just to population control, but to Africa's broader political and economic development as well."[25]

Third, they can provide opportunities for participants to ground communicative action within a life world, helping them reach a reasoned consensus centered in local agency. Current practices tend to undermine such a rationalization. As Connelly states, "The essence of population control, whether it targeted migrants, the 'unfit,' or families that seemed either too big or too small, was to make rules for other people without having to answer to them."[26]

Fourth, evaluation designs can focus on communicative action itself to determine the extent of participation, focusing on the participatory process. Did locals feel they had ample opportunities to raise issues? Were "all" issues on the table? Was every issue discussed fully and to the satisfaction of the local people? Were all validity spheres fully addressed.[27] Such communicative action evaluations can be combined with evaluations to determine audience exposure for mass media plans (who were exposed to the campaigns and at what frequency and what do they now know?); and outcomes (did the communicative experience engender reason and persuasion, leading in the long run to the desired behavior change?).

Fifth, such a prima facie western model of truth, reason and rationality is inherently relevant to a non-western context, in that the key feature of that model engages participants in "the argumentative redemption of validity claims" in every stage of the process toward reaching a consensus-bringing force of argumentative speech.[28] The contention here is that an argumentative rationalization is consistent with the very essence of Africans' rhetorical use of "a 'conversational mode,' i.e., instigating reactions by promoting 'dramas' (i.e., conflict, discussions, arguments, confrontations) as a method for internal debate . . . [and of] debates and arguments as mechanisms for the configuration of knowledge and experience."[29] The "Africa-ness" of that rhetorical dynamic is captured in a popular newspaper cartoon on dialogue on the continent:

- Speaker A: Is it true that Africans answer questions with questions?
- Speaker B: Who told you?

Finally, Habermas' cultural tradition concept bodes well for Africa's communities, which boast disparate cultural mores and values, even over small geographic areas, and invariably emphasize the centrality of culture. For example, in two southeastern Nigerian cultures—the Ngwa-Igbo and the Ubang—the body, as a physical entity, does not belong only to the individual; it is also constructed as the property of the wider community.[30] Thus, TCA will serve as a guide for crafting audience-agreed-upon, culture-sensitive campaigns for social change and for determining both their effectiveness (instrumental action) in meeting informational and motivational objectives and their reasoned communicativeness (communicative action).

It must be noted that, *prima facie*, certain realities on the continent, for example, the minuscule presence of liberal democracies, do not bode well for an application of TCA; however, its use strictly within the confines of communicating social programs for population management is both appropriate and novel. And to the degree that it engenders public interest through the pervasive use of pluralistic communicative acts in public forums and town meetings, TCA is, thus, well suited to accomplishing the purpose of this chapter.

The Theory of Communicative Action as a PCM Model in Africa

We shall now provide three sets of demonstrations of ideal speech situations that at once can potentially engender asserting and challenging validity claims and establish symmetry conditions that ensure persuasion of an actor only through the force of better, compelling and countervailing arguments, and strictly within the African context. Thus, the application of TCA emphasizes full community participation as the desideratum of making significant inroads into a subject that, for Africans, is overwhelmingly culturally sensitive. Habermas' theory advises that we make available to parties formal concepts for the objective, social, and subjective worlds and allow differentiated validity claims, all presented in simple, comprehensible predicates, existential sentences, and general propositions.

A caveat: To the degree that family size, particularly among sub-Saharan Africans, is a highly volatile, sensitive issue, it is important, then, that some scripting be agreed upon by all communication agents and adopted, even as it serves *only* as guideposts for a template for their ground activities. The point here is that a thematic framework is required for the argumentation process lest the likelihood of miscommunication, along with its knock on effects, becomes pronounced.

Reaching a shared understanding of and an agreement on campaign rationale

The first set of demonstration focuses on establishing shared agreement on a major problem in the local setting. This concerns the justification or rationale for a program intervention. Local interest and buy-in for an intervention will depend first on whether community residents understand and share with development organizations an agreement on the nature of the problem.

1. Speaker No. 1, implying a validity claim to a propositional or constative truth: *Our population is much too high for the community resources we have to sustain it.*
   Hearer A: Yes, in my division, I see hordes and hordes daylong and shrinking farmlands.
   Hearer B: No, there are so many undeveloped farmlands all over town. (Whenever a consensus cannot be reached, the alternative is to act in accordance with the persuasiveness of countervailing arguments.)
   Hearer C: Undecided (or abstain)—for the time being.
2. Speaker, implying a validity claim to a normative rightness: *Appropriate measures ought to be taken to bring our population growth in line with our existing resources.*
   Hearer A: Yes, doing nothing could exacerbate social problems.
   Hearer B: No, nothing seems awry or amiss. Why fix it, then?
   Hearer C: Undecided (or abstain)—for the time being.
3. Speaker, implying a validity claim to the subjective truthfulness, sincerity or authenticity of the symbolic expression of a speaker's feelings or of a request: *I believe the World Health Organization's (WHO) population-management program is designed not to help us, but to keep us down. I just don't trust it.*
   Hearer A: Yes, we have seen this sort of trickery before.

Hearer B: No, I believe WHO is serious and passionately committed. The planned rate seems just right for population replacement, don't you think?
Hearer C: Undecided (or abstain)—for the time being.

These initial exchanges could most productively occur in dyads that encourage the treatment of each communicative action as unique to a specific setting. There is ample room to lay out one's reasons or grounds to disagree or abstain publicly without the subject being subjected to public pressure or to public scorn in a focus-group session and, thereby, being stigmatized. And because of the homophily that exists between the cultures of the speaker and the hearer, the chances for a truly open communication geared toward reaching an agreement or consensus and mutually, symmetrically influencing attitudes and beliefs through arguments are significantly enhanced.

Because these initial dyads must quickly morph into full public exchanges that involve more community residents, the challenge here is to avoid the constraint of coercive social influence as a pressure point for behavior change—and to embrace atmospherics for argumentation and dynamics consistent with the hallmarks of the ideal speech situation.[31] This suggests a parallel to the two-step flow of mass-mediated message influence. Here, there is a two-step or multistep flow of communicative action, at two or more different levels of communicative action, that is, dyadic, then perhaps small group, then public.

Reaching a shared understanding of and an agreement on campaign themes

Communicative action thrives in a linguistic medium that is *not* necessarily used solely for transmitting information to exert influence, that is, coercion or manipulation (as in strategic action), but for knowing and reaching an understanding through language and social interaction. To be more specific, influence is acceptable if it is reached through rationally motivated agreement, that is, free and open dialogue, but not if it is achieved through deception. This means that implementing a campaign is much more than transmitting information from a population agent to an audience; rather, it is for at least two people having interpersonal relationships and reaching for an understanding. The importance of this distinction is borne out by the process-outcome dichotomy in social-scientific contexts. This means that communication, as a process, can be used to reach mutual understanding within the life world, which engenders convictions (e.g., beliefs and attitudes), precursors for behavior modification through criticizable arguments. The point here is to ensure a mutual understanding of population issues by using everyday language to make them intelligible before embarking on acts geared toward behavior change.

The PCM specialist accomplishes such a difficult task by presenting criticizable, rational expressions and by being willing to correct failed attempts; members of the audience (the hearers) may then subject each speech act to a validity claim of their own. This second set of demonstration concerns the matter of establishing shared agreement on a communication theme or themes that may be employed in addressing the local problem. This relates to the design of messages to be employed in the program. Local participation in dialogue during an intervention will depend first on whether community residents understand and share with development organizations an agreement on themes being addressed.

Again, three demonstrations:

1. Speaker, implying a validity claim to a propositional or constative truth: *Food prices and living costs in general are skyrocketing beyond our means. We should address these subjects.*
   Hearer A: Yes, and they have been for as long as I can remember.
   Hearer B: No, it depends on one's lifestyle—that is, whether one is a city or a rural dweller.
   Hearer C: Undecided (or abstain)—for the time being.

2(a) Speaker, implying a validity claim to a normative rightness: *Our tradition says that expected (or ideal) family size should be determined by our family.*
   Hearer A: Yes, family size is family business.
   Hearer B: No, family size is in the hands of Allah.
   Hearer C: Undecided (or abstain)—for the time being.

2(b) Speaker, implying a validity claim to a normative rightness: *Communicative interactions on managing our populations ought to be undertaken in a variety of spaces; therefore, this alternative:*
   Hearer A: Yes, a complementary use of several media is necessary for widespread dissemination of population news.
   Hearer B: No, one or two key local media could get it done.
   Hearer C: Undecided (or abstain)—for the time being.

2. Speaker, implying a validity claim to the authenticity or sincerity of a

symbolic expression of speaker's feelings or of a request: *I think our government's "four-is-enough" policy is designed to ensure our family's welfare.* (The Nigerian government's family planning policy is based on "four is enough"; however, any such agreement on number, if informed by Habermas' life world, requires clear cultural sensitivity on reproductive health issues.)
Hearer A: With living costs, such a guideline is meaningful.
Hearer B: No, that number is arbitrary. The government, which you support, is only trying to win votes with a clever campaign.
Hearer C: Undecided (or abstain)—for the time being.

Reaching a shared understanding of and an agreement on the properties of cultural traditions and their validity claims

Even though Habermas distinguishes communicative action from strategic action, he argues that the former complements the latter, meaning that appropriate strategic practices are permissible. Put differently, strategic action has its role, albeit minuscule, in communicative intent.

One overarching principle of communication campaigns in Africa is cultural relevance—that is, applying African historical, rhetorical antecedents as the backdrop for PCM. As noted earlier, reducing fertility is an anathema to African cultural mores; thus, population managers must consider the life world in managing population issues and help it resist a colonizing influence. How? This third set of demonstration concerns the matter of establishing shared agreement on a matter of cultural relevance that may be pertinent to the local social environment. This concerns matters of cultural sensitivity that may arise in attempts to recommend behavioral or social change in addressing a local problem.

1. Speaker, implying a validity claim to a propositional or constative truth, states, *There is a nexus in our community: our family.*
   Hearer A: Yes, it is the glue that holds us together.
   Hearer B: No, it is the bond among family members that is the real nexus.
   Hearer C: Undecided (or abstain)—for the time being.
3. Speaker, implying a validity claim to a normative rightness: *Family size ought to be determined by our cultural background and heritage.*
   Hearer A: Yes, family size should have absolutely nothing to do with our government.

Hearer B: No, size should have everything to do with overall community well-being.
Hearer C: Undecided (or abstain)—for the time being.

4. Speaker, implying a validity claim to the authenticity or sincerity of a symbolic expression of speaker's feelings or of a request: *The government believes our region should be a key target for information on birth spacing.*
Hearer A: Yes, it will, in the long run, result in limiting births through having fewer pregnancies.
Hearer B: No, the government only wants to win votes. It hasn't really evaluated the best places for information on birth spacing at all.
Hearer C: Undecided (or abstain)—for the time being.

All of the sample conversations eschew dogmatism, permit questioning, allow feedback, provide opportunities for participants to have the knowledge they need to better participate in a discourse and reach a shared understanding. Such deliberations are supported by arguments rendered fluid through permanent criticism to the degree that such a learning process can be socially institutionalized—features that are absent from current PCM programs. Improving spousal communication about fertility preferences, a major factor whose absence stymied Zambia's family planning efforts,[32] is a key tool in population management[33] through unfettered conversations to resolving some of the apathy the public ascribes to programs that limit births. In this context, such an equitable social interaction qua communicative action moves PCM to a "new" arena—a public sphere in which communities intersect and individuals interact to discuss issues of common interest and to embrace a dynamic that promotes participant inclusivity, celebrates shared values, enhances public opinion on civic solidarity and pride, arrests social marginalization, improves the accessibility of (some) views to the public sphere, creates a shared sense of ownership of a national-development issue, and discourages the rejuvenation of both social inequality and bourgeoisie qualities historically attendant on the public sphere.

Summary and Conclusion

A handful of African nations attribute their fertility transitions to national population policies, public health services, family planning programs, and improved political and economic environments. But as the two epigraphs at the beginning of this chapter indicate, fertility targets are yet to be met for the most part. Sachs argued that "we need to be working much harder to reduce the fertility rates in [the developing countries], for their benefit and for the world's."[34] To the degree that Africa has only a qualified success in stemming both its population growth rate and TFR, this chapter presents a theoretical model that can better guide population managers (the nongovernmental organizations and multilateral donors) toward ensuring that all participants become more responsible than has traditionally been the case for their own reproductive health. The application of TCA will acknowledge and de-emphasize strategic action, which guides current programs and exacerbates power inequalities such as those between governments and donor agencies, and will emphasize communicative action, which can be particularly beneficial to PCM efforts for at least three reasons.

First, participants in a communicative action are parties to "interactions in which *all* participants harmonize their individual plans of action with one another and thus pursue their illocutionary aims *without reservation.*"[35] Power inequalities between, say, campaign agents and the public, and readily associated with strategic action, are diminished by such interactions.

Second, participants are parties to making cogent arguments that could lead to negotiated compromises and agreements on whether a fertility management program is necessary in the first place; to establishing a process that approves messages and themes used in a campaign; and to raising validity claims to propositional truth, to normative rightness, to subjective authenticity, honesty or sincerity. The alignment of ideal speech acts with symmetry conditions serves as an enabler of understanding and consensus building. Thus, in every aspect of engaging in a communicative interaction for fertility program development, implementation, and evaluation, participants are encouraged to accept, challenge, contest, refute, deliberate, discuss or reject the bases of ideal speech acts and symmetry conditions. All of those possible outcomes emphasize full community participation as the desideratum of making significant inroads into a subject that, for Africans, is overwhelmingly culturally sensitive.

Third, and finally, they are parties to open, free and unconstrained conversations that frame each participant's worldviews, yet ensuring their consistency with their life world. And they are parties to interactive situations amenable to integrating the cultural tradition into the messages. Because current fertility reduction and family planning programs tend to give all of those considerations short shrift, future programs that are framed in accordance with TCA features will have a better chance to respond more effectively to Africa's population challenges.

In light of the preceding, then, it must be acknowledged here that not all strategic action should necessarily be avoided. Strategic action provides an actor with an additional choice: to consider introducing other actors to, say, a social interaction, and in specific contexts. When the context calls for an emphasis on strategic action, the outcome is manipulation; when it calls for an emphasis on a linguistically mediated interaction, the outcome is understanding.

But, more than that, strategic action in its open form tends to be significantly more democratic and audience centered than in its concealed form. Thus, to the degree that social influence through strategic action is the primary goal of population-communication managers and change agents, communicative action, then, becomes the rational action of choice to engender symmetry conditions amenable to having audiences enamored with palpably demonstrating attitudes and behaviors consistent with meeting the goals and intentions of national population policies. Habermas's reconstructive theory lends itself fully as a theoretical model for accomplishing both those actions on the ground—with much more emphasis on its communicative characteristics.[36]

Notes

1. United Nations, *World Population to 2300*. NY: UN, 2004, 115.
2. United Nations, *Millennium Development Goals Report 2010*. NY: UN, 2010a, 36.
3. Muponda, "Zimbabwe 2009 Economic Prospects." *The ZimDaily* (2009, Jan. 19). www.zimdaily.com/news/muponda27.6800.html.
4. United Nations, *Millennium Development Goals Report 2010*. NY: UN, 2010b, 12.
5. United Nations, *UNFPA Annual Report 2010*. NY: UN, 2011, 3.
6. Sachs, *Common Wealth: Economics*. NY: Penguin Press, 2008.
7. United Nations, *Millennium Development Goals Report 2009*. NY: UN, 2009.
8. Caldwell and Caldwell, 'Fertility Transition in Sub-Saharan Africa.' Paper presented in Pretoria, South Africa, 2002.
9. There are only two non-African countries in the Top Ten. They were Afghanistan ranked fourth and Yemen ranked seventh.
10. Dorius, "Global Demographic Convergence?" *Population and Development Review* 34 (2008): 519–537.
11. United Nations, *World Population Policies 2007*. NY: UN, 2008.

12. Agyei-Mensah, "Fertility Transition." *Population, Space and Place* 12 (2006): 461–477; Caldwell and Caldwell, "Fertility Transition"; Garenne, Tollman, and Kahn, "Premarital Fertility." *Studies in Family Planning* 31 (2000): 47–54; Lockwood, "Development Policy." *Journal of International Development* 7 (1995): 1–23; Onuoha and Timaeus, "Has a Fertility Transition Begun." *Journal of International Development* 7 (1995): 93–116; and Timaeus and Moultrie, "On Postponement and Birth Intervals." *Population and Development Review* 34 (2008): 483–510.

13. Lesthaeghe, Kaufmann, and Meekers, "Nuptiality Regimes." In *Reproduction and Social Organization in sub-Saharan Africa*, ed. Lesthaeghe, 238–337. Berkeley, CA: University of California Press, 1989; and Westoff and Nyblade, *Marriage and Entry into Parenthood.* Calverton, MD: Macro International, 1994.

14. Timæus and Moultrie, "On Postponement and Birth Intervals."

15. Garenne, Tollman, and Kahn, "Premarital Fertility."

16. Connelly, *Fatal Misconception*. Cambridge, MA: Belknap Press, 2008.

17. Ibid, 378.

18. Bodunde, "Techniques for Mobilizing." Paper presented in Ibadan, Nigeria, 1987.

19. Jacobstein, Bakamjian, Pile, and Wickstrom, "Fragile, Threatened, and Still Urgently Needed." *Studies in Family Planning* 40 (2009): 147–154 and Bogaarts and Sinding, "Population Policy." *Science*, 333 (2011): 574–576.

20. Habermas, *Theory of Communicative Action, Vol. 1*. Boston, MA: Beacon Press, 1984, and Habermas, *Theory of Communicative Action, Vol. 2*. Boston, MA: Beacon Press, 1987.

21. Habermas, *Moral Consciousness*. Cambridge, MA: MIT Press, 1990.

22. Pratt, "Communication Policies." *The Journal of Modern African Studies* 24 (1986a): 529–537, and Pratt, "Communicating Population." *Gazette: International Journal for Mass Communication Studies* 37 (1986b): 169–189.

23. Blake, *African Origins of Rhetoric*. NY: Routledge, 2009, and Scholtz, Braund, Hodges, Koopman and Lubben, "South African Teachers' Ability." *International Journal of Educational Development* 28 (2008): 21–34.

24. Epstein, "Talking their Way Out of Population Crisis." *The New York Times* (2011, Oct. 23): 4.

25. Ibid.

26. Connelly, *Fatal Misconception*, 378.

27. Jacobson, "Case for Quantitative Assessment." *Glocal Times* (2007, Nov.). www. glocaltimes.k3.mah.se/.

28. Habermas, *Theory of Communicative Action, Vol. 1*, 249.

29. Blanes, "Why Africans Do What They Do?" *Quaderns. Serie Monografic* 23 (2009): 123–137.

30. Izugbara, "Who Owns the Body?" *Reproductive Health Matters* 16 (2008): 159–167.

31. Habermas, in *Theory of Communicative Action, Vol. 1*, acknowledges that actual context of argumentation used to motivate participants to analyze,

discuss, and critique validity claims need not correspond fully to the ideal speech situation, which he conceives as a benchmark for criticizing deceptive and power-based communication and for measuring its outcomes.

32. Mahler, "Covert Method Use." *International Family Planning Perspectives* 25 (1999): 103–104.

33. Avong, "Religion and Fertility." *Journal of Biosocial Science* 33 (2001): 1–12.

34. Sachs, *Common Wealth*, 169.

35. Habermas, *Theory of Communicative Action, Vol. 1*, 294.

36. Habermas, *Theory of Communicative Action, Vol. 1*.

Bibliography

Agyei-Mensah, S. "Fertility Transition in Ghana: Looking Back and Looking Forward." *Population, Space and Place* 12 (2006): 461–477.

Avong, H. N. "Religion and Fertility Among the Atyap in Nigeria." *Journal of Biosocial Science* 33 (2001):1–12.

Blake, C. *The African Origins of Rhetoric*. NY: Routledge, 2009.

Blanes, R. L. "Why Africans Do What They Do: Arguments, Discussions and Religious Transmission in Angolan Pentecostal Churches in Lisbon." *Quaderns. Serie Monogràfic, 23*(2009): 123–137.

Bodunde, J., "Techniques for Mobilizing People for Effective Population Control in Nigeria." Paper presented at the Family Health/Family Planning Broadcasting Workshop, Planned Parenthood Federation of Nigeria in Ibadan, Nigeria, August-September, 1987.

Bongaarts, J., and Sinding, S. "Population Policy in Transition in the Developing World." *Science* 333(2011): 574–576.

Caldwell and Caldwell, "Fertility Transition in Sub-Saharan Africa." Paper presented at the Conference on Fertility and the Current South African Issues of Poverty, HIV/AIDs and Youth in Pretoria, South Africa, October 24, 2002.

Connelly, M. *Fatal misconception: The Struggle to Control World Population*. Cambridge, MA: The Belknap Press, 2008.

Dorius, S. F. "Global Demographic Convergence? A Reconsideration of Changing Intercountry Inequality in Fertility." *Population and Development Review* 34 (2008): 519–537.

Epstein, H. "Talking Their Way Out of a Population Crisis." *The New York Times* (Sunday Review section) (2011, Oct. 23): 4.

Garenne, M., Tollman, S., and Kahn, K. "Premarital Fertility in Rural South Africa: A Challenge to Existing Population Policy." *Studies in Family Planning* 31 (2000): 47–54.

Habermas, J. *The Theory of Communicative Action: Reason and the Rationalization of Society, Vol. 1* (T. McCarthy, Trans.). Boston, MA: Beacon Press, 1984.

———. *The Theory of Communicative Action: Lifeworld and System: A Critique of Functionalist Reason, Vol. 2* (T. McCarthy, Trans.). Boston, MA: Beacon Press, 1987.

———. *Moral Consciousness and Communicative Action.* Cambridge: MIT Press, 1990.

Izugbara, C. O. "Who Owns the Body? Indigenous African Discourses of the Body and Contemporary Sexual Rights Rhetoric." *Reproductive Health Matters* 16 (2008): 159–167.

Jacobson, T. L. "A Case for Quantitative Assessment of Participatory Communication Programs." *Glocal Times* (2007, Nov.). www.glocaltimes. k3. Mah.se/.

Jacobstein, R., Bakamjian, L., Pile, J. M., and Wickstrom, J. "Fragile, Threatened, and Still Urgently Needed: Family Planning Programs in sub-Saharan Africa." *Studies in Family Planning* 40 (2009): 147–154.

Lesthaeghe, R. J., Kaufmann, G., and Meekers, D. "The Nuptiality Regimes in sub—Saharan Africa." In *Reproduction and Social Organization in sub-Saharan Afri ca*, R. Lesthaeghe, 238–337. Berkeley, CA: University of California Press, 1989.

Lockwood, M. "Development Policy and the African Demographic Transition: Issues and Questions." *Journal of International Development* 7(1995): 1–23.

Mahler, K. "Covert Method Use Among Zambian Women is Tied to Poor Spousal Communication on Contraception." *International Family Planning Perspectives* 25 (1999): 103–104.

Muponda, G. (2009, January 19). "Zimbabwe 2009 Economic Prospects Bleak (Part 2)." *The ZimDaily* (2009, Jan. 19). www.zimdaily.com/news /muponda 27.6800.html.

Onuoha, N., and Timæus, I. M. "Has a Fertility Transition Begun in West Africa?" *Journal of International Development* 7 (1995): 93–116.

Pratt, C. B. "Communication Policies for Population Control: Nigeria in the African Context." *The Journal of Modern African Studies, 24 (1986a):* 529–537.

———. "Communicating Population Issues in sub-Saharan Africa: A Development Policy for Nigeria." *Gazette: International Journal for Mass Communication Studies, 37 (1986b):* 169–189.

Sachs, J. D. *Common Wealth: Economics for a Crowded Planet.* NY: The Penguin Press, 2008.

Scholtz, Z., Braund, M., Hodges, M., Koopman, R., and Lubben, F. "South African Teachers' Ability to Argue: The Emergence of Inclusive Argumentation." *International Journal of Educational Development* 28 (2008): 21–34.

Timæus, I. M., & Moultrie, T. A. "On Postponement and Birth Intervals." *Population and Development Review* 34 (2008): 483–510.

United Nations Department of Economic and Social Affairs. *World Population to 2300.* The Population Division, United Nations Secretariat (No. ST/ESA /SER.A /2362). NY: Author, 2004.

———. *World Population Policies 2007*. The Population Division, United Nations Secretariat (No. ST/ESA/ SER.A/272). NY: Author, 2008.

———. *The Millennium Development Goals Report 2009*. NY: Author, 2009.

———. *The Millennium Development Goals Report 2010*. NY: Author, 2010a.

United Nations Department of Economic and Social Affairs of the United Nations Secretariat. *The Millennium Development Goals Report 2010*. NY: Author, 2010b.

United Nations Population Fund. *UNFPA Annual Report 2010*. Information and External Relations Division. NY: Author, 2011.

Westoff, C. F., Blanc, A., and Nyblade, L. *Marriage and Entry into Parenthood.* (Demographic and Health Surveys Comparative Study No. 10). Calverton, MD: Macro International, 1994.

Chapter 7

Development Media: Challenges and Attempts

Genet Lakew

In recent years, Ethiopia has received international criticism for restricting press freedom, especially since it passed an antiterrorism law in 2009. This law, designed to seal the Horn of Africa country from the threat of insecurity posed by internal rebel groups, mostly independence and liberation movements, as well as neighboring Al Shabab, has been used to jail reporters and silence independence newspapers. A relevant portion of the proclamation is article 6, Encouragement of Terrorism, which states:

> Whosoever publishes or causes the publication of a statement that is likely to be understood by some or all of the members of the public to whom it is published as a direct or indirect encouragement or other inducement to them to the commission or preparation of an act of terrorism stipulated under Article 3 of this Proclamation is punishable with rigorous imprisonment from 10 to 20 years.[1]

The most recent and perhaps well-documented example is that of Swedish freelance journalists Martin Schibbye and Johan Persson. An Ethiopian court sentenced the pair to eleven years in prison for allegedly supporting terrorism after they illegally entered the Ogaden region and made contact with members of the Ogaden National Liberation Front, labeled a terrorist group by the government. While these issues appear to focus on legitimate security matters, there are others that do not and yet freedom to report on those is also restricted in the country. For instance, reporting on issues related to human rights and governance is restricted and there are severe sanctions for reporting such matters.

An in-country example of press repression in Ethiopia is that of independent newspaper editor Dawit Kebede, who went into exile in November 2011 after receiving threats of imprisonment. Just a year before in 2010, he was awarded the International Press Freedom Award by the Committee to Protect Journalists for his commitment to reporting in an insecure and hostile media environment. Kebede is only one of at least seventy-nine Ethiopian journalists forced into exile in a ten-year span since 2001.[2] Such examples illustrate the difficulty of practicing independent and critical media operations in Ethiopia. In a February 2012 United Nations News Service article, Ben Emmerson, the Special Rapporteur on counterterrorism and human rights was quoted as saying, "the anti-terrorism provisions should not be abused and need to be clearly defined in Ethiopian criminal law to ensure that they do not go counter to internationally guaranteed human rights."[3] The law has openly been criticized by human and press rights organizations like the Committee to Protect Journalists, Human Rights Watch, and Amnesty International.

Despite these shortcomings, Ethiopia has been able to use media for social and cultural purposes. In fact, Ethiopia utilizes a development journalism model as of 2008, which is a policy that "prescribes that the media and journalists should play an active role in the country's development scheme."[4] George cites Jan Voordouw, executive director of Panos Caribbean, a branch of a non-governmental organization that promotes public debate and democracy, in defining development journalism. Voordouw states that development journalism describes it as "Community journalism to achieve larger objectives—social justice, improving health, education, bringing people together."[5]

In essence, this type of journalism aims to take a closer look into deeper societal topics like religion, migration, entrepreneurship, and cultural diversity in the Ethiopian context, rather than focusing on hard news items of the day. It tries to go beyond the surface of traditional news reporting on flash points like accidents, droughts, attacks, and other similar daily events. Adoption of development journalism has not exempted Ethiopia from criticism on the grounds that this model does not leave much room for press freedom and independence. But the Ethiopian Radio and Television Agency continues to employ it as a tool to educate viewers on various topics while promoting growth and progress.

Development Media Theory

Ethiopia, unlike several other African countries, continues to practice a development media system that was widespread in the continent during the era of one party states and military dictatorships. This practice finds support in the writings of several scholars and leaders of thought during the 1960s and 1970s.[6] Denis McQuail described development media system as requiring that "certain freedoms of the media and of journalists are subordinated to their responsibility for helping in this (development) purpose. At the same time, collective ends, rather than individual freedoms, are emphasized."[7] McQuail went ahead to list several principles associated with this type of media system. These principles include the following: Media should accept to carry out its duties following national policies; that journalists have responsibilities to focus on national priorities and that the state has the right to "intervene in, or restrict, media operations, and devices of censorship, subsidy and direct control can be justified."[8]

Development media system is one of six media systems believed to exist in the world.[9] Siebert, Peterson, and Schramm had articulated four media systems in 1956 and several scholars have commented on this initial articulation before McQuail developed six systems by adding development media and the democratic-participant systems. Most African countries while continuing to focus attention on development goals have moved away from applying development media system as articulated by McQuail and instead have applied systems that have allowed more press freedoms or have used hybrids of several systems.

Communication for Development

The United Nations has provided leadership in moving away from a strict restriction on press freedom and the rights of individuals and communities to voice their opinion in the process of national development. Not long ago, various agencies of the United Nations met in Addis Ababa in 2007 to address some of the challenges in successfully implementing communication for development (C4D) throughout Africa. Communication for Development, or C4D, is the use of mass and community-based communication for the purpose of disease prevention, ecosystem protection, good governance and improved agriculture, as defined by the United Nations.[10] The round table discussion also offered recommendations on how best to alleviate some of these challenges, including placing the Millennium Development Goals in the picture. The overall consensus was to work toward a common and unified UN approach to implementing C4D strategies.

The United Nations has recognized the importance of incorporating C4D strategies into development programming that impacts not only Ethiopia but all African countries where UN agencies operate. As an influential global body, the UN's dedication to communication development is a significant advancement for the field. By the conclusion of the round table session, two major suggestions were made. The first was to consider media and communication audits and needs assessments to better understand the status of underrepresented communities and the national communication environment. Secondly, to officially form a UN inter-agency body dedicated to C4D.

Sixteen UN agencies also identified three useful avenues to consider when thinking about C4D: communication as governance or participation, sector-specific communication, and information and communication technologies (ICTs). Communication for governance endorses the idea of citizens expressing their views and taking part in public debate. It also recognizes the central role of the media in good governance by facilitating public discussion and holding governments accountable. Sector-specific information refers to tailoring a C4D strategy for different issues like gender, health, environment, and education, to that specific focus. And lastly, the round table suggested looking at innovative ways of using ICTs, making them accessible to marginalized groups, and overcoming the digital divide. Unfortunately, Ethiopia has not followed the C4D practices. While, it appears that Ethiopia has focused largely on top-down approach to development, the more effective systems have been those that allow participation at the bottom level or those that advocate bottom–up approaches.[11]

This chapter shows how McQuail's theory of development media informs development activities in Ethiopia, particularly in areas such as health and education. Additionally, it points to the fact that these development communication practices are not limited to government sources but are in fact promoted by other sources and supported by the government on conditions that they, in turn, support government policy. A downside to this development communication practice is the difficulty of applying it in such a way to take advantage of emerging media technologies or platforms such as social media. In reaction to this difficulty, the Ethiopian government had sought to either control those new media platforms or discourage them. These attempts and resistance by some Ethiopian media have predictably led to conflicts between the government and media in a number of cases as you will find in the narratives in this chapter.

Development Media

Several development media tools are used by both the Ethiopian government and non-governmental organizations and numerous groups and individuals to not only educate the public on a particular issue but bring out change in attitude and action. Development media is using radio, film, and advertisements to present a message about the harmful effects or, sometimes, benefits of a certain practice. Below we present examples of some of these uses of development media.

HIV/AIDS and Women's Health

A fitting example of using development media to bring about behavioral change is women's health and HIV/AIDS prevention organization, DKT Ethiopia. DKT is a nonprofit organization founded in 1989 that employs social marketing campaigns to promote family planning and HIV/AIDS awareness and prevention.[12] The Ethiopia branch of the organization has been active in the country since 1990. It works in collaboration with other groups like the Federal Ministry of Health and is funded by the Royal Netherlands Embassy, Irish Aid, the Department for International Development, and the United Nations Fund for Population. This is DKT's mission in utilizing social marketing methods:

> Social marketing uses traditional marketing techniques and commercial sector infrastructure to create desired product uptake and behavior changes. Socially marketed products are often subsidized to make them affordable.

All around Addis Ababa, on billboards and various pharmacy locations, condom advertisements for a brand called Sensation are strategically placed. The advertisements show the same young man and woman, presumed to be a couple, who are taking responsible steps to prevent sexually transmitted diseases. DKT Ethiopia also broadcasts forty-five second television commercials that promote the use of condoms, with a target of a young audience as is evidenced by the actors used to make the commercial. At a glance, DKT Ethiopia distributed 80 percent of all condoms in Ethiopia from 2002 to 2008. It also distributed 45 percent of oral contraceptives and 41 percent of depo provera, an injection that helps prevent pregnancy.

DKT accomplishes its mission through programs like *Wise Up*, which promotes condom use among commercial sex workers, their clients, and other at risk populations. This program operates in eighteen Ethiopian cities with activities like peer educator trainings, festivals, runs, and stakeholders' meetings. Outreach workers provide condom use demonstrations and host promotional events, including panel discussions, workshops, contests, exhibitions, and university debates. All of these activities strive to use media and mass campaign tools to reach a wide audience and effect behavioral change.

Another program DKT uses to extend its reach is Mobile Video Unit (MVU). There are three MVUs that present edutainment, education and entertainment, shows throughout various communities in Ethiopia. These shows are often held in places where large groups of people gather, such as market towns, universities, and construction sites.

On the topic of HIV/AIDS, the United Nations Development Program (UNDP) outlines a Community Conversation approach to inspire positive responses from affected communities. Community Conversation encourages participation from all sectors of a community, including men and women, young and old. It creates a dialogue so that members can comprehend the issue at hand (HIV/AIDS), ponder courses of action, and take steps forward.

> These conversations use participatory methodologies such as storytelling, active listening and strategic questioning in order to identify shared concerns and observe, question, explore and make decisions together.[13]

The goal is for citizens to share their perspectives and experiences, gain the relevant knowledge and ultimately pass on that knowledge to neighbors, family members and towns all the up to a national level. This sort of campaign is a grassroots approach to information sharing in a local and native way. Citizens are able to communicate with one another using the language and cultural context that they are most comfortable with. Of course, these conversations are led by staff well versed in a community's historical and cultural environment.

The UNDP report alludes to an Ethiopian example of Community Conversation in which Islamic religious leaders in Alaba led 130 men to get tested for HIV, inspired by these conversations. UNDP states,

> In countries where Community Conversations have been used as a primary method to involve local communities there has been a strong positive response after the science and epidemiology of HIV/AIDS was shared and understood. Communities have begun to recognize that values and actions need to change in order to combat HIV/AIDS and halt the spread of the disease.[14]

Community Conversation encourages a sense of agency among community residents and creates an open and comfortable space to discuss sensitive topics like HIV without fear and stigma from others. Instead of being told what the solutions are and what steps they need to take, residents are able to discuss among themselves and reach a conclusion. This is an empowering tool that can be used for sustainable development.

Television

An example of how Ethiopian state television uses its programming to educate its viewers and citizens is a children's animation show called *Abeba and Abebe*. It is one of the few Amharic cartoon character shows broadcast on Ethiopian Television (ETV) every weekend during children's programming hours. Each episode focuses on a specific issue, like child labor, gender equality, and education rights. These issues are fleshed out through a story involving the two main characters, siblings Abeba and Abebe, as well as their teachers, family members, and friends. Like fiction film, this television series combines education with entertainment. Because children are impressionable at an early age, this show has the ability to instill in them certain values and lessons. According to the show's website, major topics explored in *Abeba and Abebe* are: nations/nationalities; law/crime; right to health; disability rights; charity; bribery; self reliance; early marriage; social justice; environmental protection; voting/elections; information/communication; and citizenship, among many others.

In highlighting one particular episode as an example, Shewit, one of *Abeba and Abebe*'s classmates, struggles with balancing school and work in "Education First." The episode opens with their civics teacher, Malaknesh, handing back test results. Shewit arrives to class late and learns she received a failing test grade. Seeing this, Abeba suggests that they study together but Shewit nervously declines, stating that her family has for forbidden her from inviting her friends over. It is later revealed that Shewit's Aunt Aregash, who is also her legal guardian, is a hotel owner and insists that Shewit prioritize her chores at the hotel and at home over her education. Shewit is seen performing work that is too strenuous for a young girl of her size and age and she seems to be in a general sense of sadness because she is not able to study with her schoolmates or perform to her best ability in school. *Abeba and Abebe* go on a mission to highlight the legal implications of Aunt Aregash's attitude and actions toward her niece. Together with teacher Malaknesh, they state that children must not be burdened with work that is beyond their capacity in the home and especially not with work that hinders their ability to perform in school.

This "Education First" episode strives to show viewers, both children and parents or caregivers in this case, society's role in promoting and safeguarding education for youth. In this sense, *Abeba and Abebe* is an effective development media tool that can be used to educate and influence Ethiopian society. It is appealing both to children as an animated series with characters who belong to their age group or demographic but it can also be appealing to parents as a show with substance and subject matter that speaks to their role in their children's lives and in the greater society.

New Media and Development

At the time of McQuail's work on development media system, he linked most African countries to the practice of development media theory. However, his sixth theory—Democratic Participant theory—has come to fore as more individuals, non-governmental organizations, and other interest groups have used new media to promote fragmentalization of development communication. For instance, during the Arab Spring momentum of the past year or so, the power of the Internet and social media platforms like Facebook, Twitter, and YouTube to mobilize and organize the youth was evident. Young people used these sources to plan gatherings and demonstrations as well as to discuss their frustrations, goals, and opinions. Social media became a stable plane on which to obtain information and share thoughts and experiences with fellow countrymen.

But it is difficult to apply this same model of mobilization through social media avenues in Ethiopia because of limited Internet access. Tweetminster and Portland, an independent consulting firm, compiled a map of Africa showing the top twenty countries that utilize Twitter the most.[15] The map was completed by analyzing 11.5 million geo-located tweets over the course of three months in 2011. The results show that South Africa is the most active Twitter country at 5 million users, with Kenya, Nigeria, and Egypt following behind closely. Rwanda, Morocco,

Algeria, and Tunisia are in the next bracket of users, with a range between 61,000 and 745,000 users. Ethiopia is one of the lightest users of Twitter, with just 6,450 users signed on to the micro blogging social media site.

According to a 2009 report by OpenNet Initiative, an academic partnership between three institutions that studies internet censorship, only 0.4 percent, or 360,000 people, of the Ethiopian population are Internet users. This means Ethiopia has the second lowest Internet penetration rate in sub-Saharan Africa.[16] In addition, the government owned Ethiopian Telecommunications Corporation has a monopoly on providing internet access to the country.

But perhaps most problematic is the lack of open and free access to any and all websites on the World Wide Web. Independent political and development bloggers could be a useful source of information and opinion for the public if they were allowed to operate freely. Unfortunately, website that often host dissent and opposition are blocked in the country, such as blogspot.com and Nazret.com, which aggregates various Ethiopia blogs and sources. These are only two among many other political and cultural online publications that are barred from public access.

Censorship severely limits the use of mass media for development purposes in Ethiopia. A 2010 human rights report conducted by the U.S. State Department states the following:

> The government restricted access to the Internet and blocked opposition Web sites, including the sites of insurgent groups advocating violent overthrow of the government (OLF, ONLF, Ginbot 7) and several news blogs and Web sites run by opposition diaspora groups, such as Addis Neger, Nazret, Ethiopian Review, CyberEthiopia, Quatero Amharic Magazine, Tensae Ethiopia, and the Ethiopian Media Forum.[17]

Progress in development, such as education, gender equality, social services, and health are tied to government policy and thus, open to criticism. One major critic of Ethiopian government policy on development is the EthiopianReview.com, online news and opinion journal run by Washington, D.C.-based editor Elias Kifle. His website often featured harsh and direct criticism of Meles Zenawi's policies and their impact for Ethiopians in and out of the country. For this style of reporting, Kifle was tried in absentia on charges of terrorism and sentenced to life in prison.[18]

Documentary

Beyond new media, increasingly documentary filmmaking is an effective tool that is used to forge change because it combines entertainment with education. DKT Ethiopia made a short seven-minute documentary called *HIV Superhighway* designed to educate viewers on how the disease spreads through long term concurrent partnerships. It is meant to be shown in schools, anti-AIDS clubs, community organizations, and NGOs.

Ordinary individuals are also using this medium for development communication. Rahel Zegeye, an Ethiopian migrant worker who currently lives and works in Lebanon is an individual who decided to take this route. She is one of hundreds of thousands of East African women who seek domestic work in Lebanon, Saudi Arabia, United Arab Emirates, and many other Arab and Middle Eastern countries, mainly for improved economic opportunities. In the ten years that she has been in Beirut, Zegeye has experienced hardships that come with working as a domestic worker in a foreign country. She has also witnessed her fellow migrant workers get beaten, suffer illnesses, receive sexual advances by male employers and other men who live in the home, and withholding of monthly payment. Based on these experiences, Zegeye began to regret her decision to leave her home country for domestic work and realized that her employment agents had fed her false dreams.

Zegeye made a decision to educate Ethiopian women about the realities of migrant work in the Middle East and inspire them to reconsider this path. She set out to provide others the information she wishes she had ten years ago before leaving Ethiopia. Zegeye chose to advance her mission through filmmaking, a skill she was exposed to and had practiced in her former years. She received training on making short films while working with homeless communities in Ethiopia's capital city, Addis Ababa. As a domestic worker, she was only given Sundays as a rest day. Over the course of four years, Zegeye enlisted the skills and services of a cameraman and a few actresses, themselves migrant workers from Ethiopia and Sudan, to make a film called *Beirut.*[19]

Based on a collage of real stories Zegeye has experienced herself or witnessed, *Beirut* gives viewers a nuanced look at young East African migrant workers in Lebanon. It depicts a story slightly different than what is presented in the media of helpless and abused workers. *Beirut* shows women who choose to run away from their employers because the conditions were too harsh to handle. But because their legal documents are kept with the employer or hiring agent, these women become illegal immigrants with very limited employment opportunities. For this reason, they turn to prostitution for income and adopt harmful behaviors such as drinking and using recreational drugs.

Zegeye was moved to action by her negative experiences. She mobilized her pain, anger, frustration and helplessness into a media effort to urge young Ethiopian girls considering the same career path to educate themselves and understand the realities. Picking up the video camera was Zegeye's tool to stop another one of her countrywomen from ending up in the same abusive homes, dangerous streets or inhospitable country. Her goal, she says, is not to criticize or move Lebanese domestic employers, lawmakers or other members of society to change. But her aim is to place power in the hands of the young women who come to Lebanon seeking work and wealth. Zegeye believes they are the ones who can prevent the potential abuse and mistreatment they might face by choosing to remain in Ethiopia and seeking other options in their home country. That is how she believes the situation can be eased and prevented.

Zegeye has faced challenges in allowing her film to be viewed by her target audience: young women in Ethiopia. Despite not getting final approval by the Ethiopian embassy in Lebanon to release the film in her home country, Zegeye is seeking other means of showing *Beirut*. She has partnered with Migrant Workers Task Force (MWTF), a Lebanese grassroots organization that advocates for domestic workers, to screen her film and get the word out about her work. Zegeye's film has increasingly gained media attention from news networks, including the BBC, AllAfrica.com, and Tadias.com. She has also received coverage from Ethiopian news sites and blogs.

Perhaps as *Beirut* gets more public traction, the Ethiopian embassy will recognize the importance of the story and be willing to work with Zegeye in getting the film approved. Her work highlights an important and relevant issue within the Ethiopian community and Diaspora. This was further illustrated by a video that went viral of a Lebanese man, Ali Mahfouz, physically abusing and dragging an Ethiopian domestic worker on the streets of Beirut. She could be heard refusing to heed to the man's demands in Amharic and resisting entering a car. The video offset an angry sea of responses from Ethiopians in and out of the country. She was found dead a few days later inside a psychiatric hospital, reportedly having committed suicide using bed sheets. The Ethiopian community in Washington, D.C., planned a rally and vigil in front of the Lebanese embassy in March 2012 to commemorate Alem Dechasa. Such strong responses show the urgency of this topic and Zegeye's film could be a powerful tool in bringing about behavioral change among young Ethiopian women.

Conclusion

Implementing development media in Ethiopia is a challenging task because of restrictions on press freedom and communication. Ethiopian television and radio belong to the state owned Ethiopian Radio and Television Agency and thus independent broadcast mediums do not exist. But this same media enterprise has chosen to adopt a media system of development journalism since 2008, dedicating its reporting to support growth and innovation efforts in the country. This stance is not without its critics, as it is seen as interfering with media independence and press freedom.[20] Television shows, awareness commercials, and documentaries shown on Ethiopian Television (ETV) illustrate the development journalism stance of the media network.

Several programming and activities implemented by NGOs show that it is possible to independently utilize media campaigns to effect development and behavioral change in Ethiopia. These efforts are instrumental in helping the country reach some of the Millennium Development Goals related to women's health, gender equality, universal education, HIV/AIDS, among other issues. In addition, local organizations that choose to focus on these issues do not face much opposition from the government. Ethiopia's civil society law of 2009 bars foreign organizations or foreign-backed domestic organizations from working on human rights, women's rights, children's rights, disability rights, conflict resolution, and democracy and governance issues.[21] But NGOs working on development areas like health and education do not have legal funding restrictions placed on them. This makes it easier, and necessary, to focus development media campaigns on fulfilling basic health and education goals instead of human rights and governance topics.

Organizations like DKT Ethiopia, children's television programming like *Abeba and Abebe*, and individuals like Rahel Zegeye prove that development communication campaigns are effective tools at bringing about social and health behavioral change. There is room in Ethiopia's rule rigged media structure to employ helpful campaigns to provide education and assistance to citizens who need it. Such organizations and individuals just must be sure to steer clear of prohibited development efforts, as long as the civil society law and antiterrorism law of 2009 stand effective in Ethiopia. Perhaps the endorsement of C4D efforts by major world organizations like the UN will make way for a more expansive and free nonprofit work environment in the future.

Notes

1. Petre, "Ethiopian Anti-Terrorism: Limiting Human Rights." Consultancy Africa Intelligence (2012, March 16). www.consultancyafrica.com/.
2. Committee to Protect Journalists (CPJ), "Dawit Kebede Joins Ethiopia's Exiled Journalists." CPJ (2011, Nov 21). www.cpj.org/2011/11/.
3. United Nations Human Rights (UNHR), "Ethiopia: UN Experts." UNHR (2012, Feb. 2). www.ohchr.org/EN/.
4. Skjerdal, "Development Journalism Revived."*Equid Nova: African Journalism Studies* 32 (2011): 58–74.
5. George, "What is Development Journalism?" *The Guardian* (2009, Nov. 22). www.guardian.co.uk/.
6. Nkrumah, *African Journalist*. Dar-es-Salaam: Tanzania Publishing House, 1965, and Ng'weno, "The Role of the Press in a Developing Country." Paper presented at the IPI Assembly in Nairobi, Kenya, June 4, 1968.
7. McQuail, *Mass Communication*. Newbury Park, CA: Sage, 1987, 121.
8. Ibid.
9. Siebert, Peterson, and Schramm, *Four Theories of the Press*. University of Illinois Press, 1956; and Altschull, *Agents of Power*. NY: Longman, 1984..
10. United Nations (UN), "2007 UN Report." United Nations (2007). Http: // unesdoc.unesco.org/images/.
11. White, "Is 'Empowerment' the Answer?" *Gazette: The International Journal for Communication Studies* 66 (2004): 7–24; Moemeka, *Development Communication in Action*. Lanham, MD: University Press of America, 2000; and Moemeka, "Development Communication for Developing." *Gazette: The International Journal for Communication Studies* 59 (1997): 379–393.
12. DKT is an international company named after D.K. Tyagi who was a pioneer of family planning in India. Phil Harvey, a former employee of CARE International, founded DKT in 1989 in Washington, D.C. and the company operates in several countries in Africa, Asia and South America.
13. UNDP, "Communication for Development." Oslo, Norway: UNDP's Oslo Governance Centre, 2009, 37.
14. Ibid, 38.
15. Karanja, "Africa: New Research Reveals How Africa Tweets (Analysis)." Afrika.no: The Norwegian Council for Africa (2012, Jan. 26). www. afrika.no/Detailed/21139.html.
16. OpenNet Initiative, "Ethiopia: Country Report." OpenNet Initiative (2009, Sep. 30): Http://opennet.net/research/profiles/ethiopia/ and also see Onwumechili, *Reform, Organizational Players*. Lewiston, NY: Edwin Mellen Press, 2003, 113–132.

17. US State Department, "Ethiopia: 2010 Country Report on Human Rights Practices." (2010, April 8). www.State.gov/j/ drl/rls/hrrptt/2010/.

18. CPJ, "Ethiopia: Life Sentence for Blogger." CPJ (2012, Jan. 26): www.cpj.org/2012/01/Ethiopia.

19. Lakew and Meharena, "Africa: Women Filmmakers Tell Their Stories." AllAfrica.com (2012, Feb. 17). Http: // www. allafrica.com/stories/.

20. Skjerdal, "Development Journalism Revived."

21. Center for International Human Rights, *Sounding the Horn: Ethiopia's Civil Society Law Threatens Human Rights Defenders*. Northwestern University School of Law: Center for International Human Rights, November 2009.

Bibliography

Altschull, H. *Agents of Power: the Media and Public Policy*. NY: Longman, 1984.

Center for International Human Rights. *Sounding the Horn: Ethiopia's Civil Society Law Threatens Human Rights Defenders*. Northwestern University School of Law: Center for International Human Rights, November 2009.

Committee to Protect Journalists (CPJ). "Dawit Kebede Joins Ethiopia's Exiled Journalists." CPJ (2011, Nov 21). www.cpj.org/2011/11/.

———. "Ethiopia: Life Sentence for Blogger, Prison for Journalists." CPJ (2012, Jan. 26). www.cpj.org /2012 /o1/ethiopia.

George, S. "What is Development Journalism?" *The Guardian* (2009, Nov. 22). www.guardian.co.uk/.

Karanja. "Africa: New Research Reveals How Africa Tweets (Analysis)." Afrika.no: The Norwegian Council for Africa (2012, Jan. 26). www.afrika.no/Detailed/ 21139.html.

Lakew, G., and Meharena, R. "Africa: Women Filmmakers Tell Their Stories." AllAfrica.com (2012, Feb. 17). www.allafrica.com/stories/.

McQuail, D. *Mass Communication Theory: An Introduction* (2nd Ed.). Newbury Park, CA: Sage, 1987.

Moemeka, A. *Development Communication in Action: Building Understanding and Creating Participation.* Lanham, MD: University Press of America, 2000.

———. "Development Communication for Developing Societies: Facing the Realities." *Gazette: The International Journal for Communication Studies* 59 (1997): 379–393.

Ng'weno, H. "The Role of the Press in a Developing Country." Paper presented at the IPI Assembly in Nairobi, Kenya, June 4, 1968.

Nkrumah, K.. *African Journalist*. Dar-es-Salaam, Tanzania: Publishing House, 1965.

Onwumechili, C. *Reform, Organizational Players, and Technological Developments in African Telecommunications: An Update*. Lewiston, NY: Edwin Mellen Press, 2003.

OpenNet Initiative. "Ethiopia: Country Report." OpenNet Initiative (2009, Sep. 30):Http://opennet.net/research/profiles/ethiopia/.

Petre, C. "Ethiopian Anti-Terrorism: Limiting Human Rights and Democracy." Consultancy Africa Intelligence (2012, Mar. 16). Http// www. consultancyafrica.com/.

Siebert, F., Peterson, T., and Schramm, W. *The Four Theories of the Press*. University of Illinois Press, 1956.

Skjerdal, T. "Development Journalism Revived: The Case of Ethiopia." *Ecquid Nova: African Journalism Studies* 32 (2011): 58–74.

United Nations (UN). "2007 UN Report on Communication for Development." United Nations (2007). Http: //unesdoc.unesco.org/images/.

United Nations Development Program (UNDP). *Communication for Development: A Glimpse at UNDP's Practice.* Oslo, Norway: UNDP's Oslo Governance Centre, 2009.

United Nations Human Rights (UNHR). "Ethiopia: UN Experts Disturbed at Persistent Misuse of Terrorism Law to Curb Freedom of Expression." UNHR (2012, Feb. 2): www.ohchr.org/EN/.

United States Department of State. "Ethiopia: 2010 Country Report on Human Rights Practices." (2010, April 8). www.State.gov/j/drl/rls/hrrpt/2010/.

White, R. "Is 'Empowerment' the Answer?: Current Theory and Research on Development Communication." *Gazette: The International Journal for Communication Studies* 66 (2004): 7–24.

Chapter 8

Issues and Perspectives on Community Engagement for Behavior and Social Change

Ike Ndolo and Hilary Ozoh

What are the most important issues in the life of a community? As simple as this question appears to be, yet so discordant are the answers often provided by respondents. Why is there such divergence in the responses of people to this question? It is perhaps indicative of the fact that community development is often used loosely and what it signifies often viewed in several ways. In diverse countries in Africa, it becomes even more challenging, given the thousands of communities in such nations. But the difficulty in responding meaningfully to the question does not just arise because of a country's sheer size and diversity. The challenge is in trying to provide an answer that would indeed have real value in the life of the people that make up communities.

For many, the answer can be perceived from the way a community appears. This pattern of perception may account for why some have come to perceive community development as "the physical transformation of a backward community to stages represented by the symbolic presence of such structures as modern buildings or town halls, schools, hospitals, roads and bridges, pipe-borne water, electricity, etc."[1] With many communities lacking basic infrastructure for growth and development, concern for development has often been skewed in the direction of providing just basic infrastructure and equating the availability of these infrastructures with development. The onus of providing all these, traditionally, has been seen as that of the government. But it is also increasingly perceived as the responsibility of the communities themselves. Increasingly, the perception has crystalized that communities have to strongly complement government in the provision of the basic infrastructure that they require in order to live well. This is perhaps the rationale for many of the community self-help projects that abound in Africa. In many of these communities, the quest for development has often meant that the community mobilizes resources, builds and sustains the necessary organizational climate before making the required infrastructures available to residents of the community.

But this conception of community development has come to be seen as more cosmetic than substantive, yielding results or achievements that are not sustainable. Assessments of communities after provision of some of the infrastructure often show that the mere provision of basic infrastructure has not always resulted in qualitative improvement in the well-being of the people. In many of these communities, children still die unnecessarily from preventable diseases; women die needlessly during pregnancy and child birth from problems whose resolution would seem to be elementary. What has become evident is that it takes more than just infrastructure for meaningful impact to be made to the well-being of people that compose a community. This scenario draws historical foundation from the experiences and conclusions arising from the rather intensive modernization-oriented projects that had been the order of the day during an era when the interface between communication and development was biased toward the modernization paradigm. Waisbord had in a rather telling comment noted what real development is not about:

> . . . development was not restricted to just building roads, piping water and distributing electricity. Nor was it limited to efforts to increase neither farm yields nor switching farmers over to cash crops...When piped water arrived, it was frequently used for washing rather than drinking and cooking because people disliked its flavour.[2]

National leaders adopted the Millennium Declaration in September 2000 perhaps in recognition of the fact that central to all results of efforts at development is the quality of life of the people that compose a community. The adoption committed nations to a global partnership to reduce extreme poverty and setting out a series of time-bound targets with a deadline of 2015. This became known as the Millennium Development Goals (MDGs). The MDGs articulate eight goals against which nations, and ultimately the well-being of human beings that compose them, can be assessed. One of them, MDG Number 4, expressly focuses on the survival of children.

In many sub-Saharan African countries, the mortality rate of children is very high. The rate is considered against two of the very basic indicators that are used for measuring it, namely, infant mortality rate, defined as the probability (expressed as a rate per 1000 live births) of a child born in a specified year dying before reaching the age of one; and under-five mortality rate, expressed exactly in the same way, except that the probability of death is before the child reaches the age of five.

According to the 2007 Multi-Indicator Cluster Survey conducted by Nigeria's Bureau of Statistics, infant mortality rate and under-five mortality rate stands at 86 and 138 per 1,000 live births, respectively in the country.[3] Based on projections from these figures, it seems most likely that Nigeria would not meet the MDG 4 goal of reducing by two thirds its under-five mortality rate.[4] The morbidities and deaths are occurring in many communities and households.

The mortality rates are the clearest indication of whether children in any society are surviving and, therefore, whether a community is improving in terms of the human development. These then are the real issues of development that is germane to a community. These are the issues that are never supplanted by mere massive provision of infrastructure. The issues cover the different specter of the social sector, including health, education, water, sanitation, hygiene, and basic protection of children from violence, abuse and exploitation. Communities are developing on the basis of how well they are doing on the indicators that track these issues. Communities have been helped to self-track developments within their domain through a system of maintaining Community Information Boards such as the one in Table 8.1. The system, where it has been introduced, has helped communities address a number of challenges and offered communities the following participatory communication advantages:

- Increasing attention to the day-to-day wellbeing and immediate needs and rights of children and women.
- Stimulating communities to dialogue on issues, best ways to address them and the need for social data as a benchmark for ranking progress on all development objectives.
- Making it imperative to ensure active involvement and leadership of communities in tracking their own data and local information systems.
- Bringing about social change at community level with communities using public and private dialogue to work together to get what they need to improve their own lives.

COMMUNITY INFORMATION BOARD

Name of Community………………..   Year………..
Population (Adult): Male….. Female…..   # Children less 1 year old…..
Children under 5 yrs old………..   # primary schools………………
Children (6–12yrs) Male…..Female……   Children (13–15yrs)Male…Female….
Total enrolled primary Schools this yr: Male….. Female…..   # Community Improved Water resources…

S/N		Jan–Mar	Apr–Jun	Jul–Sep	Oct–Dec
1	Number children born.				
2	Number children registered at birth				
3	Number children under 1 year receiving first dose OPV at birth				
4	Number children under 5 years receiving DPT 3				
5	Number children not gaining weight				
6	Number orphans				
7	Number children attending primary school (Boys…girls…)				
8	Number households with long lasting insecticide treated nets (ITNs)				
9	Number households with latrines				
10	Number functional improved community water resources				
11	Number pregnant women attending ante-natal clinic				
12	Number women died during pregnancy or within 6 weeks of delivery				
13	Number children died within one month of birth				
14	Number children died before 5 years of age.				
15	Number community dialogue sessions				
16	Number village development association meetings				

Table 8.1: Guidelines for Community Information Boards. (Source: UNICEF Nigeria: 2008)

- Creating and improving on common understanding, stimulating local assessment, analysis, action planning, implementation and ownership of services and programs.[5]

It is in the resolution of the challenges of human development, and particularly those related to the survival of children, that the importance and centrality of communication to human development becomes most evident. An event that clearly brought this to the fore is the publication of the Lancet series on child survival. One of the points clearly discernible from the reports in the series is the centrality of communication to prospects of children surviving. An analysis of the reports shows that large scale behavior change in four key practices can reduce child mortality significantly.[6] These four essential practices are exclusive breast feeding, sleeping under insecticide treated nets, hand washing at critical times and using oral dehydration therapy to treat diarrhea. For most people in resource-poor and under-served areas to adopt these practices, large-scale communication programs will be required at the household level and since they are all "life-long" in nature, a communication approach based on dialogue, shared learning and consensus building is more likely to produce results.[7]

Why are these practices considered essential for child survival and at what level do they matter most? The argument canvassed in several fora has been that these practices do not happen at the hospitals but in households and communities. Households and communities must understand the benefits of and indeed adopt these practices. However, the idea of individuals, households and communities becoming elements of development is one that is yet to settle. But it is in this idea that the real roots of development of communities lie. Many of the critical indicators that show how well a community is doing can be located in the practices of the people that compose that community. The contention is that if development practices are adopted by communities and they routinely happen in the lives of the community members, child mortality could be drastically reduced.

The conclusion, therefore, is that the pervasive and low rate of child survival and poor development indicators are attributable to the behaviour and practices of people and communities. These include the behavior of individuals and households as well as social norms and conventions. The argument goes that to even make progress on other areas of development, one has to make significant sustainable progress in the areas of individual and community behaviour. These facts apply as much to the area of child survival as it applies to infrastructure, governance, gender and equity relations, among others.

This is why the central tenet of the application of development communication should always relate to the behaviors and practices of people and communities. There must be a clear engagement that enables communities to reflect and advance themselves through practices and norms to achieve the well-being of their people, especially children and women, who often are the most vulnerable and marginalized.

Theoretical Foundation for Community Communication Assessments

A practical field problem in communication interventions for behavior and social change has been how to assess, analyse and understand behavior patterns in communities. The demand for immediate results have always meant that careful attention has not always been paid to understanding the dynamics of the community in relation to a desired behavior. Where for instance does a community stand in relation to the adoption of such essential family practices as exclusive breast feeding, hand washing, sleeping under insecticide-treated bed nets? How does a community stand in relation to practices that facilitate violence, exploitation and abuse against children? It is, therefore, imperative that communication interventions start with the element of a thorough knowledge of the situation of a community in relation to the desired behavior. But obtaining such knowledge must be systematic rather than ad hoc. It must be founded or grounded on a theoretical approach. Several theoretical perspectives have emerged showing the interface between communication and development. Among these many perspectives, it is clear that there is strength in adopting a participatory process.

But whatever the process, the initial precedent steps for community engagement must necessarily be undertaken. These are the important steps of assessment and analysis. It is often the case that field practitioners decide on a given process and rush headlong into it. But these steps are crucial if communication is to be focused, thematic and address real problems. Communication assessment of practice patterns in relation to a desired behaviour must be anchored on theoretical foundation or foundations. It should not be treated as just another survey of demography, even though; there will always be demographic correlates of current behavior patterns. What are the elements of the two critical steps of assessment and analysis? And how does one go about addressing them? Three theoretical perspectives offer useful insights that help the assessment process. They offer good conceptual basis for the mapping of behavior patterns in relation to a desired behaviour. A good theoretical perspective that can guide field level operation is the Stages of Change Theory.[8] The theory attempts an explanation of the behavior adoption process, a five-stage process, with implication that people could be at different stages in the process of adopting a desired behavior. The five stages have been very aptly explained by McKee, Manocourt, Yoon and Carnegie.[9]

1. Precontemplation: phase in which individuals have no thought about changing their behavior because they are unaware of a problem;
2. Contemplation: phase in which individuals consider making a change;
3. Decision/determination: phase in which individuals begin making plans to change their behavior
4. Action: phase in which individuals implement their specific action plan;
5. Maintenance: phase in which individuals sustain new or desired behaviors, over time.

The stages of change theory have implication for the planning and implementation of communication interventions, especially for assessment of the patterns of practice prevalent in a given community. It enables a development communication practitioner to quickly map where a community is on a given practice, the associative characteristics and dynamics relative to those patterns of practice and consequently the kinds of insight that can help propel behaviour and social change, individual characteristics dominant at each stage and the communication strategies and channels to be used. The stages of change are a theoretical guide or framework for understanding how individuals adopt new behavior. Individual community maps, comparative maps of communities on the essential family practices and child protection practices can be produced using this theoretical foundation. Communication schools can interface with development organisations, government and other partners to undertake assessment of child care practices following this and other theoretical approaches.

A second theory, the diffusion of innovation, is anchored on social or community "responsiveness to novelty and the introduction of change."[10] Its origin dates to the seminal work on diffusion of innovation by Everett Rogers following extensive study of agricultural extension work in the USA and East Africa.[11] It is a classic study of how people in a community progressively, over time, adopt innovations. It also offers immense insights on personal influence, especially on individual and household behavior, following intermediation by opinion leaders.

Innovations have been identified by Mckee et al to include technologies, attitudes, behaviors, policies, practices, which hitherto were not known to the community.[12] The usual pattern of diffusion is that innovations are not always initially perceived positively by most people in a community. Their acceptance and adoption follows a certain trajectory of beginning small with only a few members of the community, and then generally spreading until it reaches a critical mass and becomes a widely accepted technology, supported by positive attitude, behavior or practice. The Diffusion of Innovation theory identified five stages through which people go through in becoming adopters. McKee et al provides a succinct explanation of these stages: "awareness of the innovation, interest in it, trying it out, making a decision to accept or reject, and adopting or adapting the innovation into one's daily life."[13] The theory also identified five categories of people in a community with respect to the pace of adoption of innovation; namely innovators, early acceptors, early majority, late majority, late acceptors and resistors. The theory also posits that people in the community are distributed unevenly across the adoption stages.

The diffusion of innovation perspective, posits that communication is a two-way process, rather than a simple linear process. Communication channels play a critical role in the decision-making process for adoption of innovation. The theory recognises the important role of the mass media in effecting and promoting widespread dissemination and awareness of innovations, but argues that personal influence and interpersonal communication are of immense relevance in individual adoption decisions.

The perception, receptivity and pace of diffusion are affected by several factors. These factors have been described as innovation characteristics and they comprise of the following as McKee et al. stated: [14]

- Relative advantage: how much better is the new idea or desired behavior compared to existing ones that it aims to replace?
- Complexity: how simple or difficult is the new idea or behavior to understand and adopt? How many sub-steps or actions are necessary to complete the behavior?
- Compatibility: how compatible is the desired behavior with existing practices, values and cultural norms of the intended audience?
- Trialability: how much can the desired behavior be tried out before making a commitment to act on or to adopt?
- Observability: how visible are the results and outcomes of the desired behavior to the intended audience?

The diffusion of innovation approach, like the stage of change theory, is a good tool for mapping where members of a community are on a given practice and also the associative characteristics that are critical at each stage.

A final assessment tool that we will recommend is one that enables you to understand the structures and contexts that facilitate behavior and social change of individuals and groups in a community. This tool is the Social Network and Social Support Theory. The first step is to identify or map the social networks in a community. The networks are invaluable because they are the livewire of social interactions within communities. Mapping and understanding the social networks in a community will enable development communication practitioners to take cues from how these networks influence practices of members of the community. Social networks provide significant social support to their members. It is this support that can become most impactful in communication interventions. This is why it is imperative that part of the communication assessment of a community include a mapping and analysis of the social networks. Pertinent aspects of the networks to analyse have been identified by McKee et. al.:

- Structure, i.e., size and number of members in the network;
- Interaction, i.e., the frequency of contact, strength of the bond and reciprocal nature of different contacts;
- Reciprocity, i.e., extent to which resources and support are both given and received in a network relationship;
- Density, i.e., extent to which everyone knows each other; and
- Social Support, i.e., the kinds of support provided to members such as emotional, instrumental, and informational and appraisal supports.[15]

In conclusion, it is imperative to restate that meaningful engagement of communities should begin with mapping the current situation of practice of the desired behavior by community members as well as the social networks that are in the community. This mapping may be driven by the community, or the community can be catalysed by exogenous factors. The other challenge will relate to the techniques for undertaking this mapping of practice in a community. It is at once simple to state that a lot of quantitative and qualitative methods are available to do this.

The other challenge is operational. It is imperative to always recollect that results in the area of behavior and social change need not be quick-fixed. This is important because there is always an urgency associated with delivery of program results and, therefore, the pressure to resort to whatever seems a viable short cut. The adoption of a practice takes time and a three-year time frame for the achievement of a behavior and social change, even if it is a process result (for instance, a consistent pattern of action by a community), is acceptable.

Developing Community Communication Plans

A good assessment of the current situation would provide insights on whom to focus communication interventions. This section is largely informed by the Strategic Communication for Development Planning.[16] Sometimes, communication practitioners tend to be broad in developing communications plans. While it may seem fashionable and sometimes programmatically correct to have national or state plans, it is intuitively obvious that effective communication plans are those that narrow down or narrow cast to community levels, such that communities develop and manage their own plans for behavior and social change.

After a proper assessment, the next step is the undertaking of an intensive participant analysis to determine those who are associated with the problem behavior. This pattern of analysis integrates many of the elements of the different theoretical perspectives that serve as framework for behavior and social change. It integrates the elements of individual change, especially, as it is individuals that ultimately act out behavior. It also integrates elements of social change, especially in its recognition that individuals often act out within a nexus of social relations that stretches from family to the wider community. It is in this sense that a format developed at UNICEF for the development of strategic communication for development plans becomes very helpful.

Individuals linked to the problem behavior as direct actors or in some way linked to the direct actors by exertion of influence are referred to as participants. They are normally at three levels: micro, meso and macro. Participants at the micro level are categorized as either primary, secondary, or tertiary. Primary participants refer to the individual level or the person acting out the problem behavior. For instance, within the context of issues relating to child survival, it may be referring to a mother or a caregiver, those who have direct responsibility for personal action or inaction connected to caring for the survival of the child. Secondary participants are those connected to the primary participant at the level of the family. Tertiary level participants refer to those who have influence on what people do or hold as opinions in the community. For each level of primary participants, the same set of indicative questions are asked to guide analysis and planning as well as enable comprehensive understanding of the participants. These questions include:

Who is/are the participant group(s)?
What are their current knowledge, attitudes and practices on the issue?
What are the recommended final behavior outcomes?

What are the benefits of adoption of the recommended behavior /change?
What are the barriers to the recommended behavior/change?
What are the risks?
What are the main existing facilitating factors for adopting the recommended behavior?
What are the most trusted interpersonal sources of information (CBO, FBO, health centers, schools, the hairdresser, village chief, village gossip, etc.)?
What are the most trusted mass and traditional media (television, radio, community radio, community theatre, etc.)?

The process of responding to the questions would entail collection and compilation of both primary and secondary data. As is often the case, some of these data may already be available and the challenge would be retrieval. For instance, there is an established practice of periodic Knowledge, Attitudes and Practices (KAP) studies, and all one needs to do would be to access the KAP reports. However, given the impossibility of covering all communities in a given KAP study, it may not be out of place to commission community specific studies. Funding may be a major constraint, but creative ways of overcoming this kind of operational challenge may be devised.

Another issue is who drives this process of data collection? Experiences from field work indicate a preponderance of external or exogenous drivers, but nothing stops the community becoming the driver in the study of itself. Most operational challenges can be overcome once the community is driving the process. A careful primary participant's analysis would simplify the subsequent steps of the strategic communication planning process. With the community participating meaningfully and actively, primary participants' assessment and analysis would always be an interesting process.

Of course, the key point clearly evident from all of these is that behavior and social change actions are best undertaken at community, rather than the national level. In practice, since the number of communities most times is legion, and all cannot be directly reached and engaged at once or even over a reasonable period of time, the best course of action has been to progressively reach a critical mass of communities within one planning phase. One may in practice, therefore, not be able to go the whole hog at once, but certainly one can use innovative techniques that will enable a critical mass of communities, such that the changes become self-engendering or self-propelling, moving swiftly through spatially contiguous communities.

The other levels of participant analysis are the meso and macro. In developing community level communication plans, it is often necessary to always bear in mind the associative influence externals can bring to bear on the process of behavior and social change in communities. We need to identify in every circumstance these levels of participants and program for them.

The meso participants can contribute significantly to the process, joining in the broad alliance of active field players directly carrying out activities to engender change in the target communities as well as joining in crop of advocates for the necessary policy and resource enabling environments from the macro level participants. The key indicative questions for understanding the meso level participants are:

> Who is/are the participants?
> What are their attitudes toward the promoted practice?
> What are their attitudes toward the identified marginalized categories?
> What role will they play for the promotion of the practice and in addressing equity and gender (planning, coordination, training, facilitation of participatory planning, participatory communication activities, and dialogue with the community, formative supervision, etc.)? (You can prepare a workflow for visualizing all responsibilities and interaction of these participants.)
> What are their training needs?
> What are their equipment needs (training manuals, communication material, radio equipment, etc.)?
> What are other capacity gaps in promoting the practice and in addressing equity and gender? These may include issues of motivation and leadership; authority/accountability; access/control over resources; evidence-based decision making; communication/partnership.

For the macro level participants, the questions that should guide the analysis are similar to those asked at the meso level except that on the following: What role will they play for the promotion of the practice and in addressing equity and gender (endorsement of policy, allocation of financial resources, allocation of human resources, monitoring)?

Planning Results of Communication Interventions

The nature of behavior and social change and the fact that we have to work most times with different categories of participants would often entail that we pay very careful attention to the results that we plan to achieve. This warning becomes imperative given the tendency to always think in terms of effects whenever we contemplate results for communication interventions. A proper attitude would be to look for effects in terms of the behavior and practices of the micro participants, especially the primary participants and then look for process results for all other categories of participants, all of who eventually contribute to the achievement of effect on the primary micro participants. It may then be a good step for the development communication practitioner to have a good understanding of effects and process results and be able to develop objective, verifiable behavior monitoring indicators to track their achievement.

How then do we begin to define and set results for communication interventions? The result at the end of the day is in terms of what is happening in the real lives of primary participants, in terms of what they are doing or what they are not doing. Communication for development results could be linked to use of services or behaviour, knowledge, attitudes, partners' mobilization, community participation and leadership behaviour. Results in the areas of use of services, practice of behaviour, knowledge and attitude could be set at the effects level, applying most often to the micro participants. Results in the other areas could be defined in terms of process achievements or what the other categories of participants are doing to contribute to the achievement of the planned effects among the primary participants.

The logic of the participatory frameworks for development communication also articulates the importance of the process as an end in itself and not just the substantive outcome in forms of behavior. Also, communication results eventually contribute to the achievement of a higher strategic result relating to overall well-being of the people, such as reduction in child mortality.

Communication Activities for Social Change

With participants clearly identified and understood and the planned results articulated, the next logical step would be communication activities that lead to delivery of results. A wise approach is to ensure as much participation as possible in communication activities at the community level. Another wise rule is to ensure considerable movement in the delivery of activities by participants in line with principles of sticky communication. This will facilitate greater internalization and resonance of key practice behavior benefits. Indeed, when activities are carried out at the community level, whatever their type and form, adherence to the principles of sticky communication will positively affect both the reach and effectiveness of communication activities. Sticky communication is all about the elements of language, symbol, colour, movement and music and their use in the delivery of the content of a communication. The advice is that the use of a mixture of all these elements together will make a given communication sticky and memorable.

The necessity of emphasising community-driven approaches in the identification, planning and implementation of activities comes also from historical experiences. Many communities considered mass media technologies as new forms of technology, different from their customary channels and modes of communication. They should serve to supplement rather than take over as primary and dominate the customary interpersonal methods of communication that pervade communities. Interpersonal methods like dialogues, meetings, songs, and theatre provide opportunities to identify common problems and solutions on various aspects of community lives. Their value lie not in their being instruments for transmission of messages but as avenues for exchanging views and actively participating in community decisions and actions.[17]

It is, therefore, imperative to emphasize community-driven activities. A good approach would be to ask the question suggested in the communication planning format developed at UNICEF—What community led intervention should be in place to facilitate these results? This is a question that could be answered based on experience or even much better through formal or informal research. As a corollary to this first question, the next question stresses activities that will complement and reinforce the first. It states: which interventions would you use to support the community led interventions (mass media communication campaign/ events/ marketing, etc.)?

Community Dialogue

Community dialogue is one of the approaches that give expression to the demands of participatory communication for development. It is an approach that has been used quite extensively. Its peculiar feature is that it is in sync with traditional processes of communal decision evolvement and making. The challenge of using this as a communication activity for the achievement of a planned result is that it is quite intensive and requires strenuous physical efforts. Going to scale in terms of the pace of diffusing it to several communities is also a challenge. Its principal elements of community dialogue and collective action had been most lucidly articulated by Figueora, Kincaid, Rani and Lewis.[18] However, in field operations where it has been applied, two process issues come out clear—the substantive procedure for an actual community dialogue session and the process for reporting a dialogue session. Much had been done on the outcome indicators as well as process indicators for social change. Unfortunately, the reporting of community dialogue in a coherent, systematic manner by field level operators has not been adequately addressed. It is also the same with substantive content of the procedure for a session of community dialogue. Other process challenges include setting up sessions of community dialogue in a community as well as determining and identifying participants for a community dialogue session.

Usually, community dialogue sessions take place in phases, with one phase ideally leading to the other and a good facilitator moderating carefully selected key questions. The phases are:

Phase 1: Welcome, introduction and overview.

This phase explains the purpose of the session, the procedure for conducting the session, and the role of the facilitator of the dialogue.

Phase 11: Where is the community on the behavior issue in question?

The session begins in earnest here. Participants express their understandings of the problem behavior. What are the reasons or causes for this problem behavior? Participants attempt a deep analysis of the factors responsible for this problem in the community. The essence of this phase of the session is for participants to unravel and understand the problem behavior; its ramifications within the community; and the underlying conditions producing them.

Phase III: Where do the communities want to be by way of change from the current situation?

During this phase, participants begin to discuss possible directions for change that the community should embark on. They brainstorm on what should be an ideal situation for the community, relative to the behavior issue in question. Discussions can also extend to options for actions that will enable the community to move away from the current behavior practice.

Phase IV. What will community members do, as individuals and groups, to make a difference?

At this phase, the conversation becomes more specific, in terms of what individuals and groups as members of the community must do to move away from the current practice and change their behavior.

This extends to assignment of responsibilities, monitoring and enforcement mechanisms.

Table 8.2: Community Dialogue Reporting Form

Date	
Time	
Venue	
Issues (s)	
Number of participants	
Category of participants	
Problem identified with practices of community member on the issue(s)	
Consensus on actions for addressing the problem (s)	
Persons / Organization mandated to follow up on actions	
Time frames for evaluation decided on	
Modes established for community evaluation	
Follow up on defaults	
Photographs of session	
Challenges / constraints faced during the dialogue session	
Lessons learned during the dialogue session	
IEC materials distributed at the session	
List and participants of community members that participated in the dialogue	

These steps can be applied to any issue. The important point to note, however, is that if it is properly done the session when concluded will lead to community consensus and collective action.

However, the whole community will not take part in a dialogue session. How do we then select those to participate? Field experience suggests that community dialogues are best arranged in sessions of small groups followed by large group exchanges that bring the small groups together. Usually, participants are separated into different groups with similar characteristics, e.g., women, men and youths. It is important to note that the key to sustainability of practices of behaviour is adequate representation of relevant community groups in the sessions, as well as in the large group exchanges. The community dialogue sessions could be reported on a matrix like the one in Table 8.2.

Table 8.3: Monitoring/Reporting Checklist

COMMUNITY: LGA: STATE:
NAME OF ORIENTATION OFFICER: CONTACT:

ISSUES	*Instructions for interviewer*	Response *(record on this column)*
Is there an existing community resolution or action plan for the essential family practices? (Exclusive breastfeeding, sleeping under ITNs, hand washing, use of ORS, full immunization, HIV prevention)	*Record the issues on which there is/are existing community resolutions/action plans.*	
What are the details of the resolution or action plans	*Describe it/them clearly*	
What has been done so far to implement the resolution/action plan—meetings, assignments given, work started, others	*Describe clearly*	
What significant changes have been taking place in the community since the implementation started?	*Describe clearly*	
Would you like to share your experiences with other communities?	*State clearly*	

Following Up with Communities on Indicators of Behavior and Social Change

Community engagement should not end with the carrying out of activities such as community dialogue. The community dialogue for instance would often lead to community consensus and collective action. These outputs are highly dynamic and require follow through. It may be that there are strong endogenous mechanisms for following through, such that the community can always track its performance. Yet, the reality is that many of the outputs are not followed through. This accounts for the gap in the manifest expressions for change in a community and the fact that several communities remain neck deep in traditional practices, customs and norms that negatively affect the survival and development of children.

The capacity of a community to follow through on its resolutions must be deliberately built either through endogenous systems or through assistance coming exogenously. Usually, this is where governments, NGOs and other development partners come in. They can come in with assistance and systems that support such communities. In Nigeria, for example, National Orientation Agency is in a good position to lead this role. The National Orientation Agency has a Community Mobilization Officer (COMO) in every local government area in the country. These officers have received a variety of training and participate in development programs supported by international development organizations. They are well placed to interface with different community structures and institutions for purposes of following up with these communities on their resolution and action plans, especially those relating to the essential family practices for the survival and development of children. A sample matrix for following up on communities is presented on Table 8.3.

Notes

1. Gofwen, "Community Development." *Journal of Social Work* 3 (1999): www.Dspace.unijos.edu.ng.
2. Waisbord, "Family Tree of Theories." Rockeffer Foundation (2001). www.comminit.com/pdf/.
3. Nigeria Bureau of Statistics (NBS), "Nigeria Multiple Indicator Cluster." NBS (2007): www.Nigerianstat.gov.ng/.
4. Nnakwe, "What are the MDGs." Presentation at the ACSD Workshop, Awka, June 2007.
5. UNICEF, "Child Survival." Presentation at the 2007 UNICEF WCAR Regional Programme Communication Meeting, Dakar, Senegal, 2007.
6. Ibid.
7. Ibid.
8. Proschaska, DiClemente, and Norcross, "In Search of How People Change." *American Psychologist* 47 (1992): 102–114.
9. McKee, Manocourt, Yoon, and Carnegie, *Involving People, Evolving Behaviour*. Penang: Southbound and Unicef, 2000.
10. Ibid, 19.
11. Waisbord, "Family Tree."

12. McKee et al., *Involving People.*
13. Ibid.
14. Ibid.
15. Ibid, 18.
16. The Strategic Communication for Development format was presented by Guy Scandlen at a 2011 training session organized jointly by UNICEF and the Nigerian Federal Ministry of Information.
17. Waisbord, "Family Tree."
18. Figueroa, Kincaid, Rani, and Lewis, *Communication for Social Change*. NY: Rockefeller Foundation, 2002.

Bibliography

Figueroa, M. E., Kincaid, D.L., Rani, M. and Lewis, G. *Communication for Social Change: An Integrated Model for Measuring the Process and Its Outcomes*. (Communication for Social Change Working Paper, Series No 1). NY: Rockefeller Foundation, 2002.

Gofwen, R. "Community Development: What it is. What it is not." *Nigerian Journal of Social Work* 3 (1999). www.dspace.unijos.edu.ng.

McKee, N., Manocourt, E., Yoon, C.S., and Carnegie, R. *Involving People, Evolving Behaviour*. Penang: Southbound and Unicef, 2000.

National Bureau of Statistics (NBS), "Nigeria Multiple Indicator Cluster Survey, Preliminary Report.' NBS (2007): www.Nigerianstat.gov.ng/.

Nnakwe, F. "What are the MDG? Health MDGs in Nigeria." Paper presented at the ACSD workshop for Abia and Imo States, Awka, June 18–22, 2007.

Proschaska, J.O., DiClemente., C.C., and Norcross, J.C., "In Search of How People Change: Applications to Addictive Behaviours." *American Psychologist* 47 (1992): 102–114.

UNICEF. "Child Survival: A Global Overview." Paper presented at the 2007.

———. WCAR Regional Programme Communication Meeting, Dakar, Senegal, April 2007.

———. "Guidelines for the Use of Community Information Boards." Leaflet presented to participants at the training workshop on Community Information Boards in Nigeria, 2008.

Waisbord, S., "Family Tree of Theories, Methodologies and Strategies in Development (Report prepared for the Rockfeller Foundation)." Rockefeller Foundation (2001). www.comminit.com/pdf/familytree.pdf.

Part 3

Country Cases

Chapter 9

Health Communication and Nigeria: Sickle Cell Anemia in Nollywood

Tokunbo Ojo and Jan Servaes

Since the sickle cell anemia first grabbed the world's attention in 1910, millions of children have died prematurely from it. It is an inherited disease, in which children inherit a hemoglobin "HbS" gene from their parents who are carriers of the sickle cell trait or gene (HbAS gene). It is predominantly associated with people of African descent and inhabitants of tropical climates where malaria is endemic.

Symptoms of sickle cell anemia usually manifest itself between the first six and eighteen months of a child's life. Among the earliest infant/childhood signs are the swollen hand-foot syndrome (dactylitis) and severe anemic signs like paleness and fatigue.[1] Many of the children diagnosed with sickle cell anemia do not live up to the age of five. For those that do, they regularly suffer chronic pains throughout their adolescent and adult lives. This is as a result of the distorted blood cell and hemolytic crisis in their blood stream. These chronic pains are sometimes associated with jaundice, fever and severe renal problem. Life expectancy for sickle cell anemia patients is about forty-six years.[2]

It was not until the 1940s that cases of sickle cell anemia were medically recorded in Nigeria and the West African region.[3] Since then, it has remained an unresolved public health problem in Nigeria and several other countries in the Sub-Saharan African region.[4] It is one of the major causes of morbidity and mortality in Nigeria. Approximately 24 percent of the Nigerian populations are "carriers of the mutant gene and the prevalence of sickle cell anemia is about 20 per 1000 births. This means that in Nigeria alone, about 150 000 children are born annually with sickle cell anemia."[5] This prevalence rate is one of the highest in the world.

While there is neither a standard treatment for sickle cell anemia nor a permanent cure yet, the prevalence could be reduced through low cost interventions such as counseling on genetic, prenatal screening for biomarkers, and a public health information awareness campaign. However, a random review of the health awareness messages in Nigerian media, soap operas, television drama and public billboards show that there is little public awareness of sickle cell anemia.[6] Given the importance of well-defined communication process and information dissemination in people's understanding of public health issues, this definitely creates a gap in addressing the menace of sickle cell anemia in Nigeria.

With the emergence of the home video movie tradition, popularly known as Nollywood, as an integral part of Nigeria's civic public and a forum for citizens' moral education, this gap could be partly filled by Nollywood-style movies. To demonstrate this, this paper reflects on how Nollywood-style movies have engaged in the socio-education of Nigerian population on sickle cell anemia and what could be done better in raising awareness on sickle cell anemia through Nollywood movies.

Communication Strategies for Development and Social Change

Communication strategies for development and social change can be subdivided at five levels: (a) *Behavior change communication* (BCC) (mainly interpersonal communication), (b) *Mass communication* (MC) (community media, mass media and ICTs), (c) *Advocacy communication* (AC) (interpersonal and/or mass communication), (d) *Participatory communication* (PC) (interpersonal communication and community media), and (e) *Communication for structural and sustainable social change* (CSSC) (interpersonal communication, participatory communication and mass communication). [7] Interpersonal communication and mass communication form the bulk what is being studied in the mainstream discipline of communication science. Behavior change communication is mainly concerned with short term individual changes in attitudes and behavior. It can be further subdivided in perspectives that explain individual behavior, interpersonal behavior, and community or societal behavior.

Behavioral change communication (BCC), mass communication (MC), and advocacy communication (AC), though useful, will not be able to create sustainable development. Therefore, participatory communication (PC) and communication for structural and sustainable social change (CSSC) are more concerned about long term sustained change at different levels of society. Looking at *desired* or *expected outcomes*, one could think of four broad headings: (a) approaches that attempt to *change attitudes* (through information dissemination, public relations), (b) *behavioral change* approaches (focusing on changes of individual behavior, interpersonal behavior and/or community and societal behavior); (c) *advocacy* approaches (primarily targeted at policy-makers and decision-makers at all levels and sectors of society); and (d) *communication for structural and sustainable change* approaches (which could be either top down, horizontal or bottom up). The first three approaches, though useful by themselves, are in isolation not capable of creating sustainable development. Sustainable social change can only be achieved in combination with and incorporating aspects of the wider environment that influences (and constrains) structural and sustainable change.

Nollywood and Sickle Cell Anemia

Nollywood, as a cultural agency, is one of the platforms that can be harnessed for social change and health communication in Nigeria. As a permanent fixture in the mainstream proletarian milieu (community life and culture), its systematic utilization as both communication channel and process can facilitate both short-term and long-term changes in individuals and communities when it comes to prevention and management of sickle cell anemia. The results will depend on how it is appropriated.

Though predominately in English, Yoruba and Igbo languages, Nollywood movies speak to millions of audiences across the African continent and diaspora. They grip people's hearts and minds through the portrayal of "fantasies, dreams, fears, doubts, hatreds and enthusiasm of a country (*Nigeria*) that is so large that much of the rest of the continent seems to be able to find itself in there somewhere."[8] It is a "central force of affective investment for people."[9] If appropriated as a communication process in which there is a dynamic dialogic communication between senders (health educators, for instance) and receivers (general populace, in particular sickle cell anemia patients and their families, for instance), Nollywood movies can foster consciousness required for social and self-transformation leading to better modes of intervention and prevention strategies.

Since there is no cure for sickle cell anemia, communication is key in the management and prevention strategies of sickle cell anemia. Through communication, there will be an increased understanding and better awareness of the causes, impact, management and prevention of sickle cell anemia. The appropriation of Nollywood movies, in conjunction with other community resources within a wider social-cultural environment, offers a means of achieving these goals of stimulating behavior change and creating supportive environment for those with sickle-cell anemia.

Its potential of reaching large number of people also makes it one of the viable tools to harness in communication for social change and sickle cell awareness campaign. From its humble beginning in 1992, Nollywood, as an industry, has now grown to be the world's third biggest film industry after America's Hollywood and India's Bollywood. It generates an annual revenue of about $US 300 million, and releases over 1,500 feature movies on digital video disc (DVD) and video compact disc (VCD) disks every year. These movies are regularly sold in stores and marketplaces. They are also available for rental at video clubs. In addition to these, they are a regular part of the programming staples of the Africa Magic channel of South Africa's Satellite TV Company, Multi-Choice DSTV, which has over 1.5 million subscribers in Africa, Europe and the Middle East.[10]

But these potentials of being a creative space for health promotion, awareness message and dialogic communication forum for social change have not been fully harnessed. It goes without saying that many movie producers are in the movie making business strictly for commercial interests, not for social causes. As such, blending well researched social messages in a creative way with a standardized commercial style movie that appeals to large audiences is a difficult project for many producers.

Producers conceptualize Nollywood largely as a money making institution, not as an institution for the delivery of messages for social change and identity formation. As film-maker Kunle Afolayan aptly put it: "Most of the people who are executive producers of Nigerian films are mere distributors and marketers. They have little or no knowledge about film making or what creativity is about. [For them], it's about making money and once that's done, they throw that master copy into the trash and move on to the next project."[11] Therefore, it is simply providing the public with standardized but less depth stories that they will sell, as opposed to engaging in social issues that could jeopardize their quest for profit.[12]

As a result of this mindset, only a few Nollywood movies have engaged in health-related issues and awareness campaigns in an informed way, in line with the above more 'sustainable' options for social change. Among these are Zeb Ejiro's award-winning movie *Mortal Inheritance* and Ade Solanke and Jane Thorburn's Nollywood style movie, the *Family Legacy*.[13] These movies purposefully combined entertainment with a strategic communication campaign on sickle cell anemia.

Mortal Inheritance

It is the first Nigerian home movie that tackled the issue of sickle cell anemia. It was released on video home system (VHS) tape in 1995, three years after the Nigerian home video industry started. [14] The movie, based on a true life experience, explores the ordeal of living with sickle cell anemia in Nigeria. It did this within a didactic web of a dramatic love story and cultural tradition. It was a story of a female sickle cell patient (Kemi) who has beaten the odds of dying before the age of five. Now twenty-four years old, she fell in love with Chike. But when she realized Chike was a sickle cell gene carrier (AS genotype), she broke off the relationship.

She did not want to marry a sickle cell gene carrier. Based on her own experience, she did not want to bring a child into the world and allow such a child to go through a similar ordeal of pain like her own.[15] From the scientific standpoint, if she marries Chike, her probability ratio of having a child with sickle cell anemia is one to two, i.e., a 50 percent chance.

In the same vein, Chike's mother, who was a nurse, also opposed the relationship on the ground that both Kemi and Chike were not from the same ethnic group. Kemi is Yoruba and Chike is Igbo. This subplot—Chike's mother and opposition to an interethnic marriage—highlighted the salient issue of ethnic tension within the nation of Nigeria. Although the difference in cultural and ethnic background was important to her, she was also extremely concerned that Kemi would emotionally and financially wreck her son due to Kemi's frequent "health crises" and hospitalization. The movie's subtext message is that sickle cell anemia is a deadly inheritance, and this explains the logic behind the movie's title "*Mortal Inheritance*." Kemi's "SS" genotype is considered a deadly genotype that must not match with Chike's "AS" genotype.

While the movie maintained its primary pursuit of entertainment, it did not compromise its purpose of educating the audience about sickle cell anemia. It emphasized the importance of genetic counselling before marriage. The movie's protagonist, Kemi, appeared more like a "social health worker." From the opening scene, she broke the societal culture of silence and knowledge poverty on sickle cell anemia. She informed and educated virtually every character that comes her way about the sickle cell anemia. She highlighted how the sickle cell anemia could be managed through rest, regular intake of folic acid supplements and family support. All these, the movie's overall narrative showed, contributed to her prolonged and normal life. The underlying message is that while sickle cell could affect people's lives, it cannot control people's lives if there are adequate support systems in place.

In Nigeria, there is a socio-cultural stigma attached to the sickle cell disease as well as a personal guilt factor. For instance, while some parents believe that it is a dishonourable and shameful act to have sickle cell children, many often feel that they have failed in their parental responsibilities by giving birth to sickle cell children when they could have prevented the birth of these children if they had been screened for sickle cell traits before they conceived. To this end, many do not feel comfortable to talk about sickle cell anemia. To strike the nerves of audiences and parents with these deep-rooted behavioural values, Kemi, through discussions with her friend (Chike's sister), consciously challenged this attitude. She noted that sickle cell is simply a genetic condition and there is nothing to be ashamed of. It is a reality that must be accepted.

Although the movie debunked some of the stereotypes and uninformed assumptions about sickle cell anemia, it also has its own shortcomings. Kemi, as depicted in the movie, came from a well-educated rich family. Her background and family status definitely allowed her to have good medical care and high quality of life. However, her background in the movie is not reminiscent of many sickle cell patients in Nigeria. If Kemi was placed in the same socio-economic condition that an average Nigerian family lives in, she might not have lived till the age of twenty-four. A vast majority of the population does not have access to a public health system, especially access to the high quality health care at the type of private hospital that Kemi had in the movie.

According to the data from the Federal Ministry of Health Nigeria, there is an average of 2 hospital beds per 1,000 Nigerians and 1 doctor per 30,000 people.[16] Despite being an extremely oil-rich country, Nigeria spends less than $US 5 per capita on its health care sector.[17] This figure is well below the World Bank's recommended $US 18 per capita, and

$US 34 per capita that the World Health Organization's Commission on Macroeconomic and Health recommended for basic health care services in low income countries in 2001. As at 2008, the average life expectancy rate in Nigeria is forty-eight, according to the United Nations Children's Fund (UNICEF).

In view of these facts, a social interpretation of the movie's aesthetical portrayal of Kemi's "material world" might produce a dialectical meaning, depending on the audiences' socio-economical locations. For some, it is a distraction from the embedded health-related awareness in the movie's narrative structure. For others, it is the magnetic attraction that fostered a sense of common identification with Kemi's persona, situation and problem for the audiences. As Adeoti observed, "those who purchase or rent videos sometimes do so with a view of letting off the stream of a harsh existence, especially in urban areas, with a high toll of unemployment, poverty, inadequate infrastructure, congestion and exploitation."[18]

This is the challenge of saddling social awareness messages with a leisure and commercial oriented artistic medium such as home video movies. Such extensive display of affluence unconsciously trivialized the movie's core underlying messages, despite the excellent acting of the protagonist in portraying the experience and daily realities of people with sickle cell anemia in the Nigerian context.

The Family Legacy

The challenge of balancing social responsibility with pursuit of commercial gains in processes of delivering a social awareness program without any form of trivialization was not the case with the *Family Legacy*. Produced with support from civil society organizations such as the Sickle Cell Society and National Health Services (NHS) Sickle Cell in London, UK, it is a carefully crafted public communication campaign on the sickle cell anemia. It blends theatre with persuasive educational messages that is aimed at both community and individual behavior changes.[19]

Released in London, UK, its target audiences are Nigerians and Africans in the diaspora. It explores how the birth of a child (Ayo) with sickle cell anemia changed the lives and beliefs of four generations of the same family. It dramatized the different facets of sickle cell anemia, its complication, management and the support network that is needed to minimize the pains that come with sickle cell anemia. It also showed the stress that families with sickle cell children go through, especially in catering for their children. It debunked the dominant myth that the sickle cell anemia is a satanic affliction and curse upon a family.[20]

When Ayo was diagnosed with sickle cell anemia, his maternal great-grand mother informed everybody that it was a spiritual curse on the family. The family was cursed, she claimed, by people who were envious of their success. This scene revealed an aspect of socio—cultural belief among Nigerians on sickle cell anemia. In the orthodox traditional belief systems, especially among the Yorubas in the South West areas of Nigeria and Igbos in the South East areas of the country, children who exbihit the sickle cell symptoms are unjustly stigmatized as children from the spirit world that are born to die within a specific period of time. While the Yorubas call them "Abiku," the Igbos call them "Ogbanje."

These children are believed to be wandering spirits who travel frequently between the world of the living and world of the dead. They are like "wandering spirits" that are "born only to die for sheer relish of the mischief."[21] Many of them die before the age of five. Those who do not die before their fifth year birthday, wait until "they have reached the age where they are about to marry before they die; while others choose to die some days after marriage or immediately after the birth of their first child."[22] This is presumed to be the case because these children are on the errand of mischief with the primary goal of turning their parents into wretched souls. As result of "their impish character, these fiendish *Abiku* children are happiest when their mothers are 'like Niobe, all tears' at their deaths. It is universally believed among the Yoruba people that a bereaved mother's tears are highly valued in the assembly of the *Abiku* spirits."[23]

Although there is no scientific evidence to show a linear connection between sickle cell and these belief systems, it is still a widely held belief among Nigerians, especially in the rural areas. It is not uncommon for parents and family members to seek herbalists or spiritualists to cast out the evil spell and satanic spirits in their children, as opposed to seeking medical help. In most cases, these spiritualists would ask them to appease the gods of the spirits through sacrifices. These sacrifices might include cutting of children's thumb fingers, killing of lambs and bathing in rivers at dawn, as a way to "wash away the evil spell" and appease the spirits of death.

In the last scene of the movie, the maternal great-grand mother's superstitious belief was laid to rest when genetic screening confirmed that there is a genetic trait of sickle cell anemia in the family. She was relieved of her childhood guilt of being responsible for the death of her own sister, who incidentally died of sickle cell anemia, when they were young. In her mind, she thought that her sister died because she cursed her when they had a misunderstanding.

Another misconception about the sickle cell anemia that was addressed in the movie is the issue of infidelity. In some quarters, it is believed that women who give birth to children with sickle cell anemia must have cheated on their husbands. When Ayo was diagnosed with sickle cell anemia, his father Femi instantly disowned him and accused Ayo's mother, Simi, of infidelity. "I don't have any trait like that . . . . If this child has the disease, it is not my child," Femi declared as he walked out on his wife, Simi.

The scene evoked the burden of motherhood to which many female audiences could easily relate to in the social-cultural contexts of Nigeria. It exposed and ridiculed the entire masculine traditional attitudes to motherhood and womanhood. Posed as an existential concrete situation, it stripped away the cultural covering that weaves together mothering, infidelity and sickle cell anemia under the patriarchal condition as a way to deconstruct the underlying base of Femi's action.

Implicit in Femi's action is the value-system that he cannot father a "sickler," a Nigerian derogatory name for people with sickle cell anemia, because he hardly falls sick. By this token, he believed that he could not be a carrier of sickle cell trait (HbAS gene). While in practice, he might not have high frequency of illness, it does not mean that he did not have sickle cell trait. As the studies have shown, individuals with heterozygous HbAS state (sickle-cell trait) have highly protective immunes against diseases such as malaria that is common among Nigerians. They barely fall sick, except when they are or find themselves in areas where there is a low level of oxygen. In these individuals, the differential binding of parasites to HbS versus HbA significantly reduces the possible level of host immune response which substantially ameliorates other local and systemic inflammations associated with malaria.[24] This protection can be passed genetically from parents to their children.

To illustrate this disconnection between the scientific reality and socio-cultural value system, the movie used Femi's action as basic contradictions to mediate the scientific reality around sickle cell trait and the perception of this reality held by audiences who might identify with Femi's character and action. As Freire suggested, the utilization of a certain element of contradictions allows an existential concrete situation to be posed to "people as a problem which challenges them and requires a response and not just at the intellectual level, but at the level of action."[25] After many persuasions, Femi did the genetic screening that confirmed that his son inherited sickle cell trait from him. He changed his perception and started taking actions to improve his family's life changing situation. In a nutshell, the *Family Legacy*, as a theatre of conscientization, provided a platform for the contextual analysis of high prevalence rate of sickle cell anemia among Nigerians and Africans at large. It challenged socio-cultural perceptions of sickle cell anemia and people's attitudes towards sickle cell anemia.

In addition to the short dramatic movie that debunked myths, misconceptions and superstitions about sickle cell anemia, it provided an impetus for action and social change. The movie has special feature sections that included personal accounts of a young adult living with sickle cell anemia and a narrative account of parents of a teenage girl who has sickle cell anemia. Also in these special feature sections of the movie there were (1) real-life counseling sessions during the pregnancy, (2) specialist's advice on sickle cell anemia and screening process, and (3) brief information guidelines for health professionals on how to use the *Family Legacy* movie in the health promotion campaign. In addition to this, it provided contact information and a resourceful website— www.familylegacy.org.uk—where people could get help and assistance in London, England. The clips of the movie as well as the movie's special feature section are also available on the website for public viewing.

As a problem-posing educational movie, it "regards dialogue as indispensable to the act of cognition which unveils reality."[26] To this end, in addition to being screened on Nigerian owned television networks (BEN TV) on Sky TV, and Nigeria-based HiTV that is also available on SkyTV, the *Family Legacy* is also being shown in community centers and places such as barber shops. This is to allow the movie's content to dialogue with people within their "own world" or setting. In this context, it is not just only the awareness that is important, but individuals' identification and social engagement with the sickle cell problem in such a way that there will be critical thinking and community dialogue for self-efficacy.

Self-efficacy is important not because of "reverence to individualism but because a strong sense of personal efficacy is vital for successful adaptation and change, regardless of whether they are achieved individually or by group members putting their personal capabilities to the best collective use."[27] Through a collective problem sharing approach, self-reflection, dialogue, and critical thinking, the *Family Legacy*'s goal is to make each individual a social agent in the management and prevention of sickle cell anemia.

Conclusion

As this chapter demonstrated, without compromising the pursuit of entertainment, Nollywood can be used to raise health awareness and foster people's understanding of the socio-cultural belief militating against the management of diseases such as sickle cell anemia. However, for this to happen, it requires a collaborative framework and partnership among civil society groups, government, artists, playwrights, movie producers, media outlets and members of the society. As seen, in particular with the *Family Legacy*, when there is this collaborative framework, there will not be pressure for profit in which producers see theatre as simply "theatre for entertainment," not as organic works that are committed to the social cause.

However, when engaged as the organic works for social change, they must provide a realistic understanding of the health problem being addressed. "This does not mean that each play should give a comprehensive and complete economic, political explanation, but rather should show in a simple way that each aspect of social reality forms an integrated part of a greater whole."[28] Apart from deepening awareness of the problem, ways of addressing the problem and creating social transformation must be integrated into the story plot. This obviously involves painstaking research of the issue and ways of ameliorating or solving problems. This is why collaboration and partnership are vital in the use of Nollywood for social change.

Notes

1. Miller et al., "Prediction of Adverse Outcomes," New England Journal *Medicine* 342 (2000): 83—89.
2. Infectious, "Serious Illness among Children." *Science Daily* (2007, May 4): www.sciencedaily.com/releases/2007/, and Platt et al., "Mortality in Sickle Cell." *New England Journal of Medicine* 330 (1994): 1639–1644.
3. Serjeant, "Editorial: Mortality from Sickle Cell." *British Medical Journal* 330 (2005): 432–433.
4. Olayemi, "Mortality from Sickle Cell." *British Medical Journal* 330 (2005): www.Bmj.com/content/330/7489/, and Serjeant and Ndugwa, "Sickle Cell Disease in Uganda." *East African Medical Journal* 80 (2003): 384–387.
5. WHO, "Sickle Cell Anemia," A Report by the Secretariat to the Fifty-ninth World Health Assembly A59/9 Provisional agenda item, 2006.
6. Many of the public health awareness campaigns, initiated by government and civil society groups, focus predominately on HIV/AIDS prevention, family planning, malaria and safe sex practices.

7. Servaes, *Communication for Development.* Los Angeles: Sage, 2008, and Servaes, "Harnessing the UN System." *The International Communication Gazette* 69 (2007): 483–507.

8. Barrot, "Video is the AIDS." *In Nollywood: The Video Phenomenon in Nigeria*, ed. P. Barrot, 3–9. Oxford: James Curry, 2008a.

9. Dolby, "Popular Culture and Public Space." *African Studies Review* 49 (2006): 31–47.

10. Barrot, "Audacity, Scandal and Censorship." In *Nollywood: The Video Phenomenon in Nigeria*, ed. P. Barrot, 43—52. Oxford: James Curry, 2008b. Nigerian movies constitute 80 percent of programming in Africa Magic Channel. In 2009, Multi-Choice DSTV signed a Memorandum of Understanding with Sickle Cell Foundation Nigeria, a non-governmental organization, to make the fight against sickle cell a corporate social investment project. An awareness campaign documentary, "Keeping Hope Alive" on sickle cell was released in December 2010. The documentary was to be aired on MultiChoice DSTV in Nigeria and across Africa. Sickle Cell Foundation Nigeria was founded in 1994 by Professor Olu Akinyanju, a renowned Nigeria medical researcher and former professor of medicine at the University of Lagos Teaching Hospital.

11. Otas, "Nollywood: A Touch of Class." *The African Report* 28 (2011): 83–85.

12. Adeoti, "Home Video Films." *Journal of African Cinemas* 1 (2009): 35–56.

13. Ejiro's *Mortal Inheritance* won ten of available sixteen awards at the 1996 Nigeria's Movies Award (Otherwise known as THEMA Movies Award) organized by *Fame* Magazine. The awards included the best English movies' actress of the year that went to the actress that portrayed Kemi.

14. *Living in Bondage* was the first home video movie that was released on VHS in 1992. It sold more than 750,000 copies. Its instant success marked the beginning of Nollywood as an industry and a significant cinematic shift from the celluloid films, indigenous stage story telling performance, television serials and travelling theater, which had defined the Nigerian cultural industry for more than 70 years.

15. Though at the end of the movie, Kemi married Chike and had a baby who happened not to have sickle cell anemia or SS genotype.

16. Federal Ministry, *Revised Policy—Programme and Strategic Plan of Action.* Abuja: Federal Ministry of Health, 2007. Available online: www. fmh. gov.ng/images/stories/documents/FMOH_NHMIS_Policy.pdf. (last accessed March 13, 2012).

17. Federal Republic of Nigeria, *National Human Resources for Health Strategic Plan: 2008-2012(Draft)* Abuja: Federal Ministry of Health, 2007.Available online: www.who.int/workforcealliance/countries/Nigeria_HRH StrategicPlan_2008_2012.pdf. (last accessed March 13, 2012).

18. Adeoti, "Home Video Films," 43.

19. Iyamide Thomas, Regional Care Advisor of Sickle Cell Society in London said in an interview that *Family Legacy* was created by the NHS Sickle

Cell and Thalassaeemia Screening Programme that sought to raise awareness of sickle cell testing and the disease among high risk groups such as Africans and Caribbeans. The script for *Family Legacy* was developed based on information and stories extracted from workshops and meetings with patients, families, and caregivers; and consultation with civil society groups such as Sickle Cell Society.

20. This is also confirmed in the work of Stuart J. Edelstein, renowned scientist and professor emeritus at the University of Geneva who did an ethnographical and scientific research in rural areas of Nigeria. His research findings are documented in his 1986 book *The Sickled Cell: From Myths to Molecules.*

21. Idowu, *Olodumare: God in Yoruba Belief.* London: Longmans, 1962.

22. Mobolade, "The Concept of Abiku." *African Arts* 7 (1973): 62.

23. Ibid.

24. Luzzato, Nwachuku-Jarrett, and Reddy, "Increased sickling of parasitized." *Lancet* 295 (1970): 319–322, and Cholera et al., "Impaired cytoadherence." Proceedings of the National Academy of Sciences of the United States of America 105 (2008): 991–996.

25. Freire, *Pedagogy of the Oppressed.* NY: Continuum, 2010.

26. Ibid, 83.

27. Bandura, *Self-Efficacy: The Exercise of Control.* NY: W.H. Freeman and Co, 1997.

28. Bates, "Popular Theater." *Adult Educator's Quarterly* 46 (1996): 224–236.

Bibliography

Adeoti, G. "Home Video Films and the Democratic Imperative in Contemporary Nigeria." *Journal of African Cinemas* 1(2009): 35–56.

Amaghionyeodiwe, L.A. "Government Health Care Spending and the Poor: Evidence from Nigeria." *International Journal of Social Economics* 36(2009): 220–236.

Bandura, A. *Self Efficacy: The Exercise of Control.* NY: W.H. Freeman and co, 1997.

Barrot, P. "Video is the AIDS of the Film Industry." In *Nollywood: The Video Phenomenon in Nigeria*, ed. P. Barrot, 3–9. Oxford: James Currey, 2008a.

——— "Audacity, Scandal and Censorship." In *Nollywood: The Video Phenomenon in Nigeria*, ed. P. Barrot, 43–52. Oxford: James Currey, 2008b.

Bates, R.A. "Popular Theater: A Useful Process for Adult Educators." *Adult Educators Quarterly* 46(1996): 224–236.

Cholera, R., Brittain, N. J., Gillrie, M.R., Lopera-Mesa, T. M., Diakité, S. A. S., Arie, T., Krause, M.A., Guindo, A., Tubman, A., Fujioka, H., Diallo, D. A., Doumbo, O.K., Ho, M., Wellems, T.E., Fairhurst, R. M. "Impaired Cytoadherence of Plasmodium Falciparum-infected Erythrocytes Containing Sickle

Hemoglobin." *Proceedings of the National Academy of Sciences of the United States of America* 105(2008): 991–996.

Dolby, N. "Popular Culture and Public Space in Africa: The Possibilities of Cultural Citizenship." *African Studies Review* 49(2006): 31–47.

Edelstein, S. J. *The Sickled Cell: From Myths to Molecules.* Cambridge, MA: Harvard University Press, 1986.

Federal Ministry of Health. *Revised Policy —Programme and Strategic Plan of Action.* Abuja: Federal Ministry of Health, 2007. Available online: www.fmh.gov.ng/images/stories/documents/FMOH_NHMIS_Policy.pdf (last accessed March 13, 2012).

Federal Republic of Nigeria. *National Human Resources for Health Strategic Plan: 2008-2012(Draft).* Abuja: Federal Ministry of Health, 2007. Available online: www.who.int/ workforcealliance/countries/ Nigeria_HRH Strategic Plan _2008_2012.pdf (last accessed March 13, 2012).

Freire, P. *Pedagogy of the Oppressed.* NY: Continuum, 2010.

Idowu, E. B. *Olodumare: God in Yoruba Belief.* London: Longmans, 1962.

Infectious Diseases Society of America. "Serious Illness Among Children With Sickle Cell Disease Reduced With Vaccine." *Science Daily* (2007, May 4). Http: // www. Sciencedaily.com /releases/2007/05/070501115006.htm.

Lambo, E. "Making Good Health a Real Dividend of Democracy for Nigerians under the New Administration." Notes to the members of the National Health Council, Abuja, 2003.

Luzzatto, L., Nwachuku-Jarrett, E.S., and Reddy, S. "Increased sickling of Parasitized Erythrocytes as Mechanism of Resistance against Malaria in the Sickle Cell Trait.' *Lancet* 295(1970): 319–322.

Miller, S.T., Sleeper, L.A., Pegelow, C.H., Enos L.E., Wang, W.C., Weiner S.J., Wethers, D.L., Smith, J., and Kinney, T.R. "Prediction of Adverse Outcomes in Children with Sickle Cell Disease." *New England Journal of Medicine* 342 (2000): 83–89.

Mobolade, T. "The Concept of Abiku." *African Arts* 7(1973): 62–64.

Olayemi, E. "Mortality from Sickle Cell Disease in Africa." *British Medical Journal*, 330 (2005, Feb.25). www.Bmj.com/.

Otas, B. "Nollywood: A Touch of Class." *The African Report*, 28 (2011): 83–85.

Platt, O.S., Brambilla, D.J., Rosse, W.F., Milner, P.F., Castro, O., Steinberg, M.H., and Klug, P.P. "Mortality in Sickle Cell Disease: Life Expectancy and Risk Factors for Early Death." *New England Journal of Medicine* 330 (1994): 1639–1644.

Sambo, M.N., Ejembi, C.L., Adamu, Y.M., and Aliyu, A.A. "Out-of-Pocket Health Expenditure for Under-five Illnesses in a Semi-Urban Community in Northern Nigeria," *Journal of Community Medicine & Primary Health Care* 16 (2005): 29–32.

Serjeant, G. R. "Editorial: Mortality from Sickle Cell Disease in Africa." *British Medical Journal*, Vol. 330 (2005): 432–433.

Serjeant, G. and Ndugwa, C. "Sickle Cell Disease in Uganda: A Time for Action." *East African Medical Journal*, 80 (2003): 384–387.

Servaes, J. (ed.). *Communication for Development and Social Change*. Los Angeles, CA: Sage, 2008.

———. "Harnessing the UN System into a Common Approach on Communication for Development." *The International Communication Gazette*, 69 (2007): 483–507.

WHO. "Sickle Cell Anemia." A Report by the Secretariat to the Fifty-ninth World Health Assembly A59/9 Provisional agenda item, 1996.

Chapter 10

Moving the Signpost in Cameroon? Resolving Inter-Village Rivalry through a Form of Communication

Kehbuma Langmia

The concept of development communication supposes that any meaningful transformation of the lives of a people can come about as a result of a skillful use of a communication technique. The process of cognitively changing views and attitudes about a given group of people's society constitutes development from a positive standpoint. Walter Rodney said it plainly in *How Europe Underdeveloped Africa* that western development model has failed Africa's development initiatives.[1] It was the attempt to extricate the cocoon that has warped the minds of his people for decades in colonial bondage in the Guyana in order to push in the Africa-centered development model of communal participation that created the wrath between him and the forces of darkness that caused his untimely death. To undo Western development model that privileged a select few at the expense of the masses proved to be a daunting task for Rodney and he died prematurely.

Fifteen years also have passed since Claude Ake published *Democracy and Development in Africa* in which he lamented the poor state of affairs in Africa.[2] Africa, according to him, was regressing with decreasing wealth and health for the populace, malnutrition, abject poverty, diseases, war and death. The language of development either from the economic, social, political and cultural sphere sprung out of the lips of newly independent national leaders of Africa after achieving independence. But more than forty years since most of them took power from the imperialist masters, Africa still waxes on the threshold of fundamental development initiatives intended to improve the daily social, economic and political lives of its people. The opening statement in Claude Ake book says it all:

> Three decades of preoccupation with development in Africa have yielded meager returns. African economies have been stagnating or regressing. For most Africans, real incomes are lower than they were two decades ago, health prospects are poorer, malnourishment is widespread, and infrastructure is breaking down, as are so social institutions.[3]

Ake is emphatic. His views are worthy of attention whether we agree with him or not. To him, the problem of Africa is not about development. Development was never considered as an agenda by the leaders of Africa who took over power from the 1960s. "African politics has been constituted to prevent the pursuit of development and the emergence of relevant effective development paradigms and programs."[4] In principle, scholars are doing the obvious by speaking out and beating the drum of developmental initiatives in conferences, journal articles, book chapters, debating on national and international television and writing news articles in the electronic media.

Ordinary and desperate citizens all over Africa are carrying out developmental initiatives to improve their lives but are constantly faced with the overwhelming forces of political stumbling blocks, and downright rejection for no apparent logical reasons. They seem to readily welcome outside developmental initiatives because they are less threatening to the powers that be. This sometimes creates a clash of interests. Two forces are at play here: western developmental model are colliding with indigenous Africa-driven ones. Western-driven developmental model opines that most initiatives are failing in Africa primarily because they crave to follow some of the tenets of the dependency theoretical model. The idea is that what succeeds in the Western world will eventually succeed in Africa. Or to put it mildly, what succeeds in another African country can succeed in our own without due regard to cultural, religious and topographical differences. Whatever the case, dependency theory has roots in the imperialist world. This theory has various approaches to it. Scholars like Raul Prebisch believe that Western development does not really trickle down to a multiplier effect in developing countries rather they give room to poverty and destitution.[5] On the other hand, Sunkel warns that western development or what is referred to as "external influences" have an effect on developing countries.[6] This is what makes them dependent. This is a limiting factor, according to most scholars, and is responsible for the inertia or lack of growth and progress with respect to political, economic, social and cultural developments in Africa. We should imitate what works elsewhere in Europe and America and then apply it to our various countries. Part of this is strongly driven by colonial mentality as confirmed by Mazrui in his views on dependency theory:

> It is Africa's painful process of cultural westernization without technical modernization. Africa borrowed the wrong things from the West - even the wrong components of capitalism. We borrowed the profit motive but not the entrepreneurial spirit. We borrowed the acquisitive appetites of capitalism but not the creative risk-taking. We are at home with western gadgets but are bewildered by western workshops. We wear the wristwatch but refuse to watch it for the culture of punctuality. We have learnt to parade in display but not to drill in discipline. The west's consumption patterns have arrived, but not necessarily the west's technique of production.[7]

The Western "technique of production," according to Mazrui, is the cankerworm eating Africa. A borrower is often called upon to return the borrowed article to its owner. That is Africa's fate. Maybe we should consider what Chaly Sawere has called Ali Mazrui's Pax-Africana stance in defense of neo dependency theory (Africa's self-discipline, self-pacification and self-development).[8] We intend to demonstrate how external developmental push to improve lives of village people in the South West province of Cameroon started on a good note but ended on a bad one. The initiative failed to sustain itself because local people and their traditional mode of operation was not the prime initiator. Outsiders came to advise them on what to do. It was a mild form of rebuke!

Background Information

Cameroon is situated in the Western part of Africa. It is often called the miniature of Africa and the only country in Africa with English and French as the official languages of the country. There are ten provinces in Cameroon with South West province being one of them. The South West province is one of the only two provinces in the country to house the only English-speaking population of Cameroon, usually referred to as Anglophones. The capital of the South West province is Buea. The other cities in the South West province include Kumba, Tiko and Limbe. Kumba is the region where the development initiative took place.

I see the need to present the effort of ordinary Cameroonian citizens who embarked on a developmental project to resolve the conflict of three rivalry villagers. They used various communication techniques under the platform of what they termed "theater for integrated rural development" to help the villagers build a bridge over a river that connects their three villages. These villagers have been rivals for decades and they have not inter-married, shared market spaces or allowed their children to school together for this long. For three consecutive days, these experts wanted to achieve what the villagers have not achieved for decades.

This chapter examines the tenets of participatory development communication initiatives as opined by Jody Waters in what she terms the top down versus the bottom up approach to development.[9] The top down approach must not be confused with power politics in Cameroon in influencing that development project. Rather it should be understood to mean non-local citizens coming in together to convince the local people on the need for development. Some scholars, including me, have preferred to use the term "outside-in approach view of development." In this paper, I intend to chronicle the attempt by western ideologically driven experts to bring developmental initiatives to three small villages in Cameroon using what they termed "theater for integrated rural development techniques." The reason why this fits under the rubric of development communication is that they employed various communication forms like group communication, interpersonal communication, and non-verbal communication strategies in their methods to achieve their goals. Their intention to change the minds of villagers to see beyond their mental horizon and make peace with neighboring villagers constitutes development by itself. The fact that this emanates from Western ideological developmental model (outside-in approach) is debatable as we shall learn.

Kumba Development Project

The villages of Konye, Krume and Kake are situated in the South West province of Cameroon. These villages have a river that runs through their various villages and for years they have not had the courage to build a bridge across that river because they are suspicious of each other. They believe that if they build a bridge across the river then enemies from neighboring villages will cross over and cause harm to them. This animosity among the villagers has been going on for decades and that has affected economic, social and political development in the villages.

It was, therefore, critical to resolve this rooted conflict among the communities in order to make a bridge building project possible. Resource persons from Cameroon led by Professor Hansel Ndumbe Eyoh and other specialists of theater for integrated rural development in Cameroon, Africa, Europe and the United States of America travelled to Cameroon to either take part in person, advice or disburse funds for such an initiative. They believed that they could use the technique of popular theater for development to change the lives of the people of the three villages mentioned above. It should be mentioned that the power of theater, be it elite or popular, relies mostly on verbal and nonverbal communication techniques. The intersection of body kinesics and vocal gesticulations constitutes nonverbal and verbal communication strategies. This can be achieved when meaning is transmitted through the encoding and decoding processes. The encoding and decoding process of communication is done using language. In the case of Kumba, the language was Pidgin English (i.e., creolization of English) and local languages. Since this was a classic outside-in project, there was the supposition that non-local project members could speak and understand the local citizens. If that was impossible, then translation and interpretation was going to be inevitable and that is what happened in Kumba. The vast majority of the members of the project who were non-Cameroonians could not speak the local languages. They were not natives of the villages. They were outsiders and so could only speak English and/or French.

These three villages have been rivals for decades over the construction of a bridge over River Moungo that runs through their villages. This river that runs through the three villages of Konye, Kurume and Kake has a hammock in the name of a bridge. It has become so dangerous to cross the river that many inhabitants in the three villages have died in the process. But they have been unable to come together and contribute money to construct a concrete cement bridge because of acrimony that has long existed between them. The Cameroon government has been reluctant to set aside money to construct a bridge for the villagers in the midst of the conflict. These villages are the English-speaking parts of Cameroon. The Anglophones are in the minority and so micro projects of this magnitude need a political heavyweight most probably from this part of the country to fight for its realization, but the government in Cameroon, since independence has been dominated by elites from the French-speaking part of the country.

As a result, the problem remained unresolved until Dr. Hansel Ndumbe Eyoh, a professor of English at the Yaounde University and an expert in theater for integrated rural development persuaded his colleagues from around the world to join him in resolving the conflict. His colleagues had engaged in similar ventures in Kenya, Zimbabwe, Nigeria, Botswana and Malawi.[10] Their activities involved using appropriate verbal and nonverbal language techniques through the genre of popular theater to communicate to the people. They also exploited communication forms like interpersonal communication (house-to-house and person-to-person clarion call for assembly meetings), group communication (interaction with village councilors and chiefs and group theatrical display). For two weeks, all the experts were involved in meeting with the villagers and coming up with an improvised fictional play that captures their problem so they can all act it out before the villagers.

Outside-in and Inside-out Development Model

Outside-in model of development comes into play when the source is non-indigenous. Scholars like Waters would gladly refer to them as top-bottom approach.[11] Inside-out approach is when the people who are targets or recipients of the development take upon themselves to fashion out a strategy to deal with their problem and inform the larger community or the world. Waters will also call this the bottom-up approach.[12]

Was the Kumba communication development initiative a combination of local expertise model and Western model or a domination of the Western model over the local model of development? One other argument could be that the indigenous inside-out approach failed because the people stayed in acrimony with each other for long and could not foster the energy, the will and enthusiasm to get over their ego-centricism and think globally for the common good. Was success attributable to Eyoh's work using the outside-in approach? What happened to the African communal fellowship, good neighborliness that has become the trump card for espousing the African development model?

Mind you, Eyoh's team worked in collaboration with students from the Kumba local community school who were Cameroonians from the South West province of Cameroon. Were the students from the local community development center in Kumba being trained to use western model of development or local Cameroon development model that takes into consideration linguistic, cultural and ethnic impediments? Or both?

It is notable that the project recorded a short span of success while the experts were still in Cameroon. When they left, the project collapsed. The fact that the project enjoyed a short span of success indicates that the role of the local students from the Kumba community development school were not of any meaningful help. They went along with the Western development model that made the project to end in fiasco. The project failed woefully because the people timidly agreed to resolve their age-old problem to avoid embarrassment from the guests. The deep-seated problem among them was beyond the purview of outsiders. They had only a glimpse of what ailed them. That is why when they left they went back to their usual routine for minding their own business.

Dependency Theory

Dependency theorists like Sunkel, following in the footsteps of Raul Prebisch and Walter Rodney, believe that "external influences" have been the cause of Africa's developmental demise.[13] They hold the view that European models imported to Africa have failed Africa because they have made the continent to abandon its own creative model to look to the West for direction. This has not worked because we are constantly waiting for what works abroad to try it in Africa. It is basically a top down approach to development. Granted, popular theater for integrated rural development has witnessed some success stories in Nigeria and Zambia.[14] In spite of the apparent success of Nwansa's workshop to stem the tide of HIV/AIDS in Zambia, this is what he said about some of the setbacks: "the project was catalyzed with external financial support and has continued to depend largely on external donations."[15] This external funding is a trap for dependency. You play by my rules else funding will be cut can be heard from the subtext of the agreement before funds are disbursed. For fear of blame, development experts play by the grantor's rules so that when failure becomes inevitable, their argument could be "but we followed your model of development." In the Kumba project, the nine participants were supported by Swedish International development Authority (SIDA), and other donors like International Cooperative Alliance (ICA), United Nations Educational, Scientific, and Cultural Organization (UNESCO), International Development Research Centre (IDRC) and Oxfam Canada speaks volume. In fact, IDRC dispatched "two of its program officers, Dr. Sidiki Coulibaly and Dr. Anne Bernard to Yaounde to discuss the details with the coordinator."[16] Thus, this relationship was already a powerful signal. It justifies the dependency paradigm from the top-bottom perspective. Money is power. He who brings in the money dictates the tune on how it has to be spent.

Implementing change

The villagers had series of developmental problems. But the main problem, according to the outside experts, was refusal by all three villages to construct a bridge across the river. The contention was that each village chief wanted the piece of farmland not to be affected by the bridge construction. Constructing a bridge across river Moungo implied that trucks carrying heavy merchandise from neighboring towns and cities will use their farmland as transportation routes. This will result in ecological destruction of not only their farmlands but their animals. Worse still, they will have to cope with the influx of gangs and banditry that will disrupt the peace and tranquility that they have enjoyed for years. Secondly, it will make members of rival villages to gain easy access to their farmlands and other property.

After listening to the plights of the villagers, the experts decided to meet each village chief and counselors but they faced a new resistance. Each chief or counselor felt undervalued because the experts had met them after meeting others. Thus, the order of the meetings was considered critical. Furthermore, each chief or counselor feared that gossips may have transpired in meetings with others and thus making the position of the chief or counselor vulnerable. The only safety valve was to look for a way to gather all the villagers together and discuss the matter head-on. The ten members of the team listened to all three villages and discussed their long standing problems of refusing to construct the bridge.

After that, the team members decided to dramatize an improvised fictional play to the villagers. The drama was based on a story of three brothers who had three pieces of land as a token from their father who was dead. One of the brothers wanted to build his new house and wanted a truck to transport materials to that his piece of land. However, before the truck could reach his property, it had to pass through the land of one of his brothers. All the two brothers declined to allow the truck access for fear of destruction of their property. They decided to act this play and did not create a solution but instead turned to the villagers for their opinions on what they think is the solution. To their surprise, the villagers recognized that they have mirrored their longstanding problems in this story.

Through the various communicative acts (interpersonal and group), the villagers agreed to sit among themselves and resolve their problems. Before the visitors could leave after two weeks of intensive work with the villagers, they had community development meetings and the villagers agreed to contribute money towards the construction of the bridge. There was excitement on the part of the development communication experts. They were ten who traveled to Cameroon for this project. However, one of them in the name of Abubakar Toure left prematurely. Eyoh, unhappy with Toure's decision, made the following assertion:

> . . . Who knows what would have been the outcome of the workshop had Abubakar Toure not left with his benign indifference to the whole process? Kumba worked. It worked for the students of the community development specialization training school as much as it worked for the village communities.[17]

One is then left with the conviction that the fictional drama and the various forms of communication between the visitors and the natives were successful. Thanks to theater for integrated rural development, the three villages of Kaka, Kurume and Konye appeared on the road to prosperity and peace. Not so fast! Ten years later, the same expert who initiated the workshop and had been invited to other African countries to try the experiment came back to Kumba and was surprised. No bridge was constructed and the acrimony between the villagers remained the same. This is what he said:

> I asked myself where the empowerment we thought we had achieved ten years ago was? . . . I looked for one of those who had taken an active part in the workshop in 1984 and asked him to accompany me to the bridge. His amusement to the idea of the bridge did not alert me to the site that awaited me. My disappointment was total. Ten years later, there was no bridge. Rather the canes that were used to construct the hammock gaped at me not in their exquisite waves and webs but in a cacophony of cut edges.[18]

From a socio-psychological communicative view point, communal effort by the natives of Kake, Kurume and Konye by themselves could have yielded a much more plausible outcome. However, it did not and instead it raised the question: What happened?

Discussion/Conclusion

The presence of an outside force and more importantly highly educated experts from various universities in Africa and Europe created an uneasy relationship. From a nonverbal communicative standpoint, their body language especially as they watched their plight being transformed into a fictional play, created for them by outsiders, had an unpleasant feeling even though they took part in it. It was as if someone had to come and teach you how to clean your own house. You pride yourself by living in filth. Proof of it: the actors did not conclude the play but turned to the villagers to ask for their input on resolving the conflict and they quickly, without hesitation, said they knew the story was about their long standing rivalry about constructing the bridge across river Moungo. They came to self-realization. But that realization was imposed on them from the outside. Their dignity was now in question. Seen from another lens, it may have been more successful if the villagers created their own scenario to look for ways to resolve the conflict. Alternatively, success may have also come if their kith and kin, living in the cities, created the scenario. Moreover, the use of a village customary conflict resolution procedure may have been more effective than a fictional play mirroring the plight of the villagers.

Those other options reflect an alternative to top bottom solution when bottom up solution fails to yield any meaningful lasting dividend. Maybe we have to think seriously about implementing neo-dependency theoretical model espoused by Ali Mazrui in what he called Pax-Africana i.e., Africans' self-discipline, self-pacification and self-development as already discussed above. This is what can propel such an initiative forward giving the fact that it gives room for introspection and soul searching solutions. The way to empower a people to resolve a problem is not to put the face of a foreigner forward. It is not to fictionalize a serious issue in three or four days or even a week. According to Mazrui, Pax-Africana is achievable if only we are ready to allow the potential of indigenous citizens to bourgeon. There are several methods for achieving this, the most important being a thorough feasibility studies with both local people and visitors. Then meeting groups can be formed and headed by local people themselves through some directives from expert visitors without visitor presence being felt physically! Most people are apprehensive of outsiders, to say the least, and this was the Achilles' heel of the Kumba development project. If this was done systematically then money for the bridge construction could have been donated and constructed through an agreeable format that will not disrupt villagers' lives.

The efforts of visitors only enjoyed short term success. The end result was failure to achieve the desired long term goal of Kumba community development project which was aimed at what Waters called "participatory models." Rather the "orthodoxy approaches," i.e., the top down models of development was allowed to burgeon.

Notes

1. Rodney, *How Europe Underdeveloped Africa* (Washington, DC: Howard University Press: 1982) xii.
2. Ake, *Democracy and Development in Africa*, Washington, DC: The Brookings Institution, 1996, 1.
3. Ibid, 1.
4. Ibid.
5. Ferraro, "Dependency Theory" in *The Development Economics Reader*, ed. Secondi (London: Routledge, 2008), 58–64.
6. Sunkel, "National Development Policy," *The Journal of Developmental Studies* 6 (1969): 23–33.
7. Gomia, "Mobilizing the Hordes" (Cameroon: Yaounde University, 2009).
8. Sawere, *The Multiple Mazrui* (NY: Society for the Study of Islamic Philosophy and Science, 1992).
9. Waters, "Power and Praxis" in *Redeveloping Communication for Social Change*, ed. Wilkins (Lanham, MD: Rowman & Littlefield, 2000), 89–102.
10. Eyoh, "Popular Theater Revisited" in *New Theater in Francophone and Anglophone Africa*, ed. Fuck (Amsterdam: Rodopi, 1999), 95–138.
11. Waters, "Power and Praxis."
12. Ibid.
13. Sunkel, "National Development Policy."
14. Eyoh, "Popular Theater," 95–138 and Nwansa, "Theater for Development."
15. Nwansa, "Theater for Development" 16.
16. There were ten participants that began the project but one of them left before the project completion.
17. Eyoh, "Popular Theater," 13.
18. Ibid, 15. and Gomia, "Mobilizing the Hordes," 88.

Bibliography

Ake, C. *Democracy and Development in Africa*. Washington, D.C.: The Brookings Institution, 1996.

Eyoh, H. "Popular Theater Revisited" in *New Theater in Francophone and Anglophone Africa*, ed. Fuck, A., 95–38. Amsterdam: Rodopi, 1999.

Ferraro, V. "Dependency Theory: An Introduction" in *The Development Economics Reader*, ed. Secondi, G., 58–64. London: Routledge, 2008.

Gomia, V. "Mobilizing the Hordes: Radio Drama as Development Theater in Anglophone Africa." PhD diss., Yaounde University, 2009.

Nwansa, D. "Theater for Development: A Multiform Approach to Behavior Change and Empowerment in the Fight against spread of HIV/AIDS in Defense Forces." www.unesco.org/education/ uie/pdf/Mwansa.pdf.

Rodney, W. *How Europe Underdeveloped Africa.* Washington, DC: Howard University Press, 1982.

Sawere, C. *The Multiple Mazrui: Scholar, Ideologue, Philosopher and Artist.* NY: Society for the study of Islamic Philosophy and Science, 1992.

Sunkel, O. "National Development Policy and External Dependence in Latin America." *The Journal of Developmental Studies* 6 (1969): 23–33.

Waters, J. "Power and Praxis in Development Communication Discourse and Method" in *Redeveloping Communication for Social Change,* ed. Wilkins, K., 89–102. Lanham, MD: Rowman & Littlefield, 2000.

Chapter 11

Communicating Peace: Women Transcend Cultural Boundaries in Post-Genocide Rwanda

Eddah Mutua and Ayantu Tibeso

In post-conflict settings, most efforts to bring communities together tend to look at the role of communication, culture and women in the reconciliation and reconstruction process. Communication and culture have an interdependent and reciprocal relationship that manifests itself in the way we talk, act and interact with others.[1] On the other hand, United Nations continues to recognize women as core to development efforts. Secretary General Ban Ki-Moon has stated that women are integral to the peace-making; peacemaking and peace building efforts and not an after-thought.[2] This chapter seeks to advance scholarship and models of inquiry that generate knowledge about communication processes in post-genocide Rwanda. We use experiences of women survivors of the genocide to highlight factors that have shaped their communication for peace in post-genocide Rwanda. Specifically, we seek to understand the impact of women's embodiment of the post-genocide life on Rwandan cultural institutions and how the experience offers new insights to conceptualization of post-genocide peace communication. We want to highlight peace activities and emergent critical discourses which have become valuable tools to dismantle patriarchal practices barring women as key players in peace processes. Of great interest to us is the construction of the very notion of peace in context of the post-genocide Rwanda—how it is articulated and meaning assigned from the experience of violence.

This chapter discusses women's efforts to transcend cultural boundaries in post-genocide Rwanda. We do so by highlighting ways that women locate their resistance against cultural institutions that limit their role as peace makers.

We begin our discussion by describing in brief the juxtaposed position of women as insiders and outsiders in the general scheme of Rwanda culture. Next, we examine the context in which women's peace activities occur. Theoretical perspectives employed to understand and account for women's efforts to communicate peace draw from African scholars whose works attempt to add value to African knowledge models as relevant to interpreting the African reality. We believe that stepping outside of the Western knowledge to advance the study of communication processes in Africa is imperative. Therefore, Rwandan women's communication activities for peace will be examined by utilizing African cultural perspectives. We conclude by highlighting lessons learned about the intersection of gender, culture, communication, peace and development in post-genocide Rwanda.

Women as Insiders and Outsiders in Cultural Institutions

Rwandan women's embodiment of the post-genocide life serves to illustrate the complex intersection of gender, culture, post-genocide life, peace, and development. The intricate interweaving of these variables is what shapes how peace communication is perceived and performed by women. The process to validate the role of women as key players in cultural institutions mandated to create harmony and manage conflicts in society has been transformative. It has demanded the need for cultural change that is working to redefine gender roles, and giving women voice to articulate their experiences as central to the post-genocide peace building process. There are examples in literature that attest to this transformation. For example, Powley offers the importance of active participation of women in strengthening governance and legislation of laws that protect women and girls.[3] She notes that the genocide was gendered and so must be the recovery. Thus the historical and cultural understanding of gender roles as defined by division of labor has to change in post-genocide setting. As will be discussed later, the acknowledgment of roles, and experiences of women in relation to the peace-building process in post-genocide Rwanda is now part of the reconstruction process. Having said this, it is important to point out that attention on Rwandan women in peace-building processes is not just about the fact that women and children are most predisposed during any violence. It is more about the collective response by women to challenge cultural practices that were (and are still) not inclusive to women. Women's willingness to become key players in the post-genocide peace process places resistance against gender exclusion in Rwandan cultural institutions. The act of transcending culture to become active in *Gacaca* courts and formal leadership structures has enabled women to redefine their role as outsiders to insiders of the Rwandan culture. Gacaca courts are Rwandan traditional court system used for conflict resolution.

It is important that readers do not interpret this characterization of women as a way to romanticize Rwandan culture. Uwineza and Pearson observe that Rwandan women have been empowered and disempowered by aspects of indigenous culture.[4] This chapter illustrates how the post-genocide life has enabled women to transcend cultural boundaries that disempowered them from playing significant role as peace makers.

Context

The post-genocide life forms the context that allows us to explain women's communication activities. It is a context marked by devastating effects of the 1994 genocide and subsequently defines how peace communication is facilitated. We are attentive to decisions taken, processes involved and commitments made to empower women to engage in meaningful communication in a context dramatically transformed by violence.

Effects of the genocide include loss of life, widowhood, anger, distrust, fear, health issues including HIV/AIDS infection, displacement, land issues, poverty, and trauma, among others. The magnitude of the genocide in the 100 days it happened was extremely violent. Verwimp documents the scale, geography and pace of the genocide that resulted in the death of more than 10 percent of the overall population and about 75 percent of Tutsis killed.[5]

Similarly, Cohen et al. show the impact of the genocide on women. They report 250,000 women raped during the genocide and those intentionally infected with HIV.[6] Additionally, they discuss how the convergence of the genocide trauma, widowhood, poverty, and lack of education make women more susceptible to multiple traumatic events which increase the likelihood to experience posttraumatic stress disorder.

While the genocide forms the backdrop of women's peace activities, our intent is not to belabor the effects. Instead, we argue for recognition of women's efforts to overcome these atrocities through numerous ways to communicate peace. Specifically, we explore ways that women have used their experiences of the genocide and utilized culture as a resource to transcend cultural boundaries to communicate peace. Even as Rwandan women strive to find peace from violence, their actions offer lessons about how local people are capable of transforming circumstances without necessarily privileging the role of the government or international actors in the reconciliation process in Rwanda. The complexity and controversies about determining approaches to reconciliation in post-genocide Rwanda cannot be ignored. Betts calls our attention to the critical question about where approaches to post-genocide justice and reconciliation should be determined globally, nationally or locally.[7] It is, therefore, important to interrogate the role each level of approaches to reconciliation in question play in the interest of the community reconstruction in post-genocide Rwanda. But so, too, is understanding ways the post-genocide context has allowed women to determine specific approaches to reconciliation relevant to their experiences, immediate environment and vision for the future of Rwanda. This is significant given the fact that Rwandan women are seldom celebrated as "trailblazers" in using culture to redistribute power in cultural institutions traditionally reserved for men. In Rwanda, the paradox of culture in empowering and disempowering women is real. Nonetheless, women have positioned culture as a timeless resource to support the reconstruction of Rwanda. It is against this background that we derive meanings assigned to ways that women have responded to the post-genocide context. By so doing, we create a knowledge base to theorize emergent peace activities politically enacted by women to communicate peace.

Theoretical Perspectives and Literature Review

This section discusses theoretical perspectives that offer analytical and interpretive meanings assigned to Rwandan women's peace communication initiatives. In particular, the theoretical perspectives privilege African cultural knowledge as relevant to understanding the human condition in Africa. We do not wish to generalize African culture but want to highlight those universal values that apply to most Africans. Moemeka argues that "there are irrefutable fundamental factors of social order that are uniquely African, that obtain everywhere on the continent, and that have an overwhelming impact on the communicative behavior of Africans anywhere."[8] This understanding sets the context in which to examine women's peace activities in post-genocide Rwanda and their associations with perceptions of culture as a valuable resource in national peace building efforts.

The rationale for theoretical perspectives grounded on African knowledge is an attempt to decolonize the study of Africa communication. We want to contribute knowledge to ongoing scholarly efforts to privilege African knowledge as relevant to the advancement of theory development in post-conflict peace communication grounded in realities of the African communication environment.[9]

Development Communication Theory in African Context

Preference to work within broad African cultural values frameworks rather than traditional Western theories that define peace communication and/or development communication allows for the interpretation of Rwandan women's experiences to be situated within an African reality. African cultural-based framework clarifies the relevance of women's peace activities in transforming the conditions of the post-genocide setting because the activities are grounded in the local culture. How then do we view women's work in post-genocide Rwanda as transformative and one that stands to contribute to an expanded understanding of development communication in post-conflict societies?

Development communication is an area within the discipline of communication that is well theorized. For example, *Communication for Social Change*, edited by Gumucio-Dagron and Thomas Tufte, provides the most important texts in the field.[10] The anthology testifies to the diversity and depth of theorizing in this field, capturing voices from Latin America to Africa and Asia. We are persuaded to contextualize, in appropriate theories, Rwandan women's peace activities in view of their contribution to development in a post-genocide setting.

Moemeka asserts that "development communication is the application of the processes of communication to development process."[11] This perspective of development communication presumes open-ended approach, and plurality of communication and development processes which give participants the freedom to work with resources available and are considered relevant to improving the wellbeing of society. Awareness of issues to be addressed and the appropriate response needed facilitates the desired change in society. It is in this light that Rwandan women's peace activities are seen as a catalyst for peace and development.

In examining communication and development processes in post-genocide Rwanda, we choose to de-emphasize the traditional development communication models as debates in the early 1970s and 1980s have challenged their propositions. Instead, we employ culturally responsive theoretical perspectives embedded in African cultural values and bear relevance to interpreting the Rwandan experience.

Several African scholars across multiple academic disciplines offer scholarly perspectives deemed relevant to understanding the role of culture in most Africa societies. For example, John Mbiti, Frank Ugboajah, Andrew Moemeka, Kwame Gyekye, Ngugi Wa Thiong'o, Chinua Achebe among others theorize elements of African culture and their influence on peoples' way of life. Specifically, we find the works of Andrew Moemeka and Kwame Gyekye relevant in problematizing African culture and evaluation of elements of the culture that give meaning to African way of life.[12] In post-genocide Rwanda, the local culture has given meaning to ways women participate in the reconstruction of Rwanda.

The Place of African Cultural Values

Kwame Gyekye provides a framework that offers insights into interpreting the relevance of African cultural values and practices in advancing the quality of life. It is a framework for interpretation and assessment of the place of African cultural values in a scheme of African modernity.[13]

Gyekye does not present modernity as was the case in early development communication studies. He considers most African cultural values "when the appropriate and necessary amendments and refinements have been made, to be relevant to African modernity, that is, to be the cultural life of the African people in the modern world."[14] Specifically, this intellectual framework delineates positive features of African culture that can be accommodated in attaining goals and vision of the present. Some of the features relevant to our discussion include: i) communalism which is expressed in the sharing of a common social life, commitment to the common good of the community, appreciation of mutual obligations, caring for others, interdependence and solidarity; ii) human rights as upholding the dignity of all people; and iii) knowledge and wisdom whose purpose is to promote and ensure human wellbeing.[15] Gyekye's framework clarifies the position that the interpretation of African cultural values in modern Africa is contextualized in a scheme of African modernity and not western modernity.

Overall, this framework offers insights to understand and fully account for cultural shifts in the post-genocide setting that have motivated women to act for the attainment of a peaceful life in Rwanda. As a matter of fact, the post-genocide setting has necessitated the need to "amend" and "refine" local cultural values and practices to promote peace and co-existence among Rwandans. Women have drawn from local Rwandan culture to support and engage in peace initiatives and communication processes responsive to the needs of the post-genocide setting.

Communication in Communalistic Societies

Communication in African communities is largely a community affair. Moemeka observes that in African cultures, "communicative acts are engaged in fundamentally to confirm, solidify, and promote communal social order."[16] He further observes that community needs take precedence over individual needs and aspirations in communalistic African societies. In this regard, it means that different forms of communication exist to serve the needs of the community. However, this does not mean that the community subjugates the individual but rather, as Moemeka argues, the needs of the individual and those of the community "merge in a holistic manner."[17] "Individuals are an integral part of the whole—the community—and derive their place in the context of the community."[18] After all, the existence and survival of the community depends on individuals. Likewise, the individual derives a sense of self through his or her community. While individuals have special responsibilities in the community, the community is also accountable for their wellbeing. For instance, community members are responsible for caring for those who fall on bad times (sickness, death, etc.). In view of this, it is not a surprise that Rwandan women have taken to supporting each other and their communities in addressing the effects of the genocide. In essence, women's communication efforts are to be seen as an attempt to rebuild their community image, "which is critical to whom they are."[19] Thus, Rwandan women's efforts to communicate peace as survivors of atrocious violence reveal the essence of their actions to rebuild their community.

Drawing on Culture to Communicate Peace

The theoretical frameworks presented may seem to some as merely reductionist and overly simplistic given the complexity of culture in ways it defines the role of women in peace building. However, the argument at hand is not to romanticize and/or demonize culture but rather to call attention to how often we neglect to see culture as dynamic, not a stable and constant entity. Perhaps, it is now time to start thinking about culture as what humans make of it. In the discussion to follow, we will illustrate ways that Rwandan women teach humans the capability of dictating whether or not culture should change, be amended/ refined or stay the same.

We chose to highlight two cultural institutions, perhaps not exhaustive, but in our assessment resonate well with women's decisions to use some aspects of the Rwandan culture as resource to transcend cultural repression. For Rwandan women, this is an impetus to transform cultural institutions and create an identity for themselves as key actors in post-genocide peace communication. Women are constructing a new cultural narrative that justifies and locates them at the center of Rwanda's new development strategies to promote peace in the post-genocide setting.

Women's Participation in Gacaca Courts

The structure and functions of *Gacaca* courts in the post-genocide setting stand to illustrate ways that cultural values and practices have been "amended" and "refined" to include women as key players in the peace building process. Historically, *Gacaca* courts represented an indigenous model of interaction that offered opportunities for only male community members to participate in resolving community issues. Uwineza and Pearson assert that:

> Women were not allowed to participate directly in public deliberations; rather, they were expected to play an indirect role in the customary system of justice, known as *gacaca*. Through their personal networks, women would lobby their husbands and influence court decisions by proxy. Women's absence from public proceedings such as *gacaca* court hearings is evidenced by the word for witness, *umugabo*, which translated literally means "a man." Only recently did the term for witness change from *umugabo* to *umuhamya* in order to make the term gender-neutral and formally include women witnesses. Further, with *gacaca* courts revived in recent years and adapted to deal with genocide crimes, at least 30 percent of judges in modern *gacaca* courts are women.[20]

This passage reveals cultural practices that barred women access to *Gacaca* courts and how the post-genocide setting has caused the cultural shift to allow women the right to participate in the courts' activities. Women are transgressing exclusionary cultural practices and asserting themselves as active actors in the peace building process. The opportunity for women to disrupt rigid cultural institutions was welcomed in an interview with a female *Gacaca* court judge and survivor of sexual violence who stated that:

> You cannot seek justice and peace alone. We have to be together both men and women. Women have the truth and know about what happened during the genocide. We have to be heard because it is important for our own healing and bringing our people together.

Therefore, the move to include women as participants and leaders in *Gacaca* courts gives them a voice that deconstructs the representation of women beyond the superficial image of passive actors in peace activities. The plurality of voices and experiences locates women as custodians of knowledge and agents of reconciliation efforts in Rwanda. The post-genocide life as embodied by women centers their experiences as important knowledge base of what they say and do. Scott observes that "when experience is taken as the origin of knowledge, the vision of the individual subject . . . becomes the bedrock of evidence on which explanation is built."[21] Therefore, Rwandan women are taking on a political project to counteract the dominance of culture and make their experiences visible.

Women's Leadership

In Rwanda, history of women's leadership offers insights to the role women leaders can play in spearheading peace. Women's leadership facilitates the rethinking of history and culture in redefining their role as leaders in the post-genocide setting. Looking at the current positioning of women leaders in comparison to the pre-genocide era, one sees a patriarchal society that has experienced a shift in gender roles. For example, in the pre-genocide era, there were a handful of women leaders. The most cherished and celebrated woman leader was the late Agathe Uwilingiyimana, recognized for championing the peace movement. As a leading woman reformer who served in the Hutu government as Prime Minister of Rwanda from 1993 to 1994, she pushed to foster peace between warring Hutu and Tutsi political groups.

In the post-genocide era, the number of women leaders has increased drastically. The experience of genocide has caused women to "demasculinize" traditional cultural and power structures to reclaim access to leadership positions. As a result, we are seeing change at a magnitude never witnessed before genocide. The face of leadership in the historical and cultural sense has been transformed by the dark history of Rwanda. Remarkably, the transformation has been very significant with women not only occupying prime and central spaces within community and national spheres, but its impact on culture and tradition has been extensive. Nzegwu notes that the number of women represented in Rwandan legislature jumped from 5 percent in the pre-genocide era to 50 percent in the post-genocide period.[22] McCrummen wrote an article for the *Washington Post* after the 2008 elections reporting Rwanda's parliament as the first in the world where women claim the majority or 56 percent including the speaker's chair, a third of all cabinet positions held by women including the foreign minister, education minister, Supreme Court Chief Justice and the police commissioner.[23] A 2000 United Nations Development Fund for Women (UNIFEM) report *Women at the Peace Table: Making a Difference* recognizes Rwandan women's grassroots groups for their dedication to reconciliation and peace. The increased visibility of women as leaders reaffirms their recognition and place as key actors in national reconciliation and reconstruction efforts in post-genocide Rwanda.

Women Communicating Peace

In discussing women communicating peace, we examine ways that peace activities discussed above contribute to the process of promoting peace in post-genocide Rwanda. The discussion focuses on explaining ways that women's participation in cultural institutions that previously marginalized them has now become an avenue to communicate peace.

Women at Gacaca Courts as Peace Communication

Gacaca courts serve as a cultural institution that women have transcended to earn their place as key actors in the peace process in post-genocide Rwanda. Women's active participation during *Gacaca* sessions as witnesses, judges, victims, and community members is an opportunity to present women as peacemakers and custodians of knowledge about advancing the understanding of post-conflict peace communication. Women's words and actions are indicative of selfless sacrifices made by them for the sake of peace in their communities. Rather than thinking of themselves as individuals in confronting their experiences of the genocide, women have chosen to frame their desires for Rwanda in ways that favor community. The notion of community building as a strategy for peace communication in post-genocide Rwanda is exemplified in two ways. Firstly, the willingness by women to give accounts of what happened during the genocide, especially in regards to sexual violence, is admirable. Women's testimonies at *Gacaca* courts about rape, a topic that is taboo in many African traditional societies, has become a recognizable flagship for peace in Rwanda. Secondly, women's plea to the perpetrators who sexually abused them to confess the truth and seek forgiveness is a way to involve them in the peace communication process. The fact that women are able to facilitate communication between victims and offenders in *Gacaca* courts gives promise to national reconciliation process.

It is safe to note that the different roles that women play in *Gacaca* courts are geared at communicating peace. Women are judges, leaders, witnesses, victims, peacemakers, community members, citizens, mothers, widows, victims of HIV/AIDS, caregivers among others. These roles intersect with cultural values and experiences of the genocide to allow women to become a viable voice of justice, and vision for peaceful co-existence in post-genocide Rwanda. Importantly, the males and society has come to accept these roles.

Women's Leadership as Peace Communication

Leadership is one aspect of the Rwandan culture where women reveal the relevance of culture as a resource to promote peace communication. In the post-genocide era women recognize the need to take on leadership that is rooted in the cultural fabric of the country. Women's leadership activities define the effort to reclaim their traditional role as leaders in the family. As Izabiliza has noted, historically, Rwandan women played an important role in the country's governance through the institution of the queen mother.[24] While colonialism degraded the role of women as leaders, the post-genocide setting has caused women to strive to lead at the higher echelons of society.

Demonstration of women's peaceful leadership styles is evident in diverse peace initiatives organized by women to promote peace and national reconciliation. Here are some examples:

Peace Baskets Project

This is a project often cited by Rwandan leaders and citizens as a demonstration of women's peaceful leadership styles. The project brings together widows of genocide victims and wives of the killers—genocidaires—to weave peace baskets. The act of weaving the baskets is interpreted as symbol of unity, weaving communities together and paving way for well-knit communities in the post-genocide era. The project facilitates communication between the women which symbolizes how far women are willing to go to make peace in their country. The enthusiasm to talk with each other is an attempt to restore trust, innocence, empathy and compassion that were taken away by the genocide. This is an act of communication that counteracts the narrative of violence on women by women documented in *Rwanda: Not So Innocent, When Women Become Killers.*[25] Instead, the project among many others advances a narrative of women's inevitability as peacemakers in post-genocide Rwanda. It reveals the integrity of Rwandan women in their efforts to make peace in their communities.

Ishyirahamwe Abaruhuje Women's Group

The loose English translation of this group's name is "people who have come together." It was formed in 1995 to bring together genocide widows, women whose husbands were in prison for genocide crimes or had fled their homes, women who had returned from exile and genocide survivors residing in Butare town. Diversity of the membership reflects the very purpose of the women coming together, namely; unity, reconciliation, forgiveness, community building and peace.

Association de Solidarite' des Femmes Rwandaise (ASOFERWA)

The English translation is Rwandan Women Association for Solidarity. The association was started by a group of educated Rwandan women to offer support to orphans, widows, HIV/AIDS infected women, child-headed households, the elderly, minors in detention centers, and traumatized women and girls. ASOFERWA works with community members to impact economic empowerment and peace education in several regions in the country including Kigali, Ntarama and Kamonyi districts. Other aspects of the association's programming include counseling, civic education such as encouraging women to participate in *Gacaca* court sessions, and education for girls and women. These activities facilitate communication and interactions that send a strong message about the need for Rwandans to act collectively, and stand in solidarity to address effects of the genocide. This group provides an opportunity for vulnerable women to bounce back and rebuild their country.

The Forum of Women Parliamentarians (FWP)

Rwandan parliament marks the record number of women parliamentarians in the world. No other parliaments, whether in the developed Europe or in countries such as the United States that stresses equality and freedom have such number of women sat in the national legislative house. The legislative forum offers opportunity for women parliamentarians to advance Rwandan women's leadership and chart ways to champion women's needs at the legislative and policy level of the government. Women face the burden of violence and it matters how laws and policies are enacted and implemented to enable women to make significant contribution to national reconciliation and nation building. The women parliamentarians join in purpose to advocate for women's needs and gender equality within the provision of the law and recognition of local cultural values. For example, by building upon good cultural traditions, FWP was able to seek support from male parliamentarians to pass the Gender Based Violence (GBV) bill. Recognizing the role of culture as a resource to advance women's leadership gives a new face to efforts in place to rebuild Rwanda.

Implications for Development Communication Theory and Future Research

In what ways does this chapter contribute to theorization of development communication in post-genocide settings?

It was noted at the beginning of this chapter that there was a need to grow a knowledge base about peace communication in post-conflict settings in Africa. Indeed the paucity of research in this field limits our knowledge about how we can share experiences and best practices. Thus, the imperative of reflecting on lessons learned about Rwandan women in this chapter is to show ways that women's peace activities have become avenues to improve the lives of many Rwandans. Some of the lessons learned are about how to conceptualize, implement and sustain communication processes and activities in post-conflict situations.

We identify two lessons that will define the success of these processes and activities. The success lies in considerations made for the long-term goal to attain and sustain processes geared at achieving post-genocide peace. Firstly, women's peace activities point to the longevity of their commitment (long-term) to rebuild Rwanda. Sustainability of peace is evident in ways that the idea to transform the post-genocide setting is conceptualized from within, not from outside Rwanda. Women's understanding of their circumstances renders their efforts to a lifetime commitment to rebuild lives, communities and the country. Furthermore, women have chosen to work together to realize this goal. They demonstrate that an effort which is truly participatory and grounded in the positive cultural values of a people is indeed genuine and can lead post-genocide Rwanda to greater new heights. In contrast to the dominant top-down approaches, the experiences of Rwandan women exemplify the sustainability of bottom-up approaches which privilege local voices. Communication for Development scholars, largely influenced by the work of Paulo Freire, have long recognized the importance of genuine community involvement in sustaining social change. At the core of participatory engagement is the realization that people are agents of change, and not merely patrons of top-down projects, as such, empowerment and community ownership are critical to social change processes which seek long-term solutions.[26]

Secondly, it is important to bear in mind that the circumstances that changed Rwanda came about in unconventional ways. It is "undesirable catalyst" that caused and still bears unimaginable human suffering to Rwandans. It is for this reason that the life-transforming activities spearheaded by women are different from experiences elsewhere in Africa about communication for change. In post-genocide Rwanda, specific communication for change/ transformation of society is desired. In view of this, there is a need to pay attention to approaches considered appropriate to counteract the change brought about by the genocide and to attain the desired change which is peaceful co-existence among Rwandans. The Rwandan women have formulated and articulated their response for the desired change from within in order to advance actions that enhance human life in a post-genocide setting.

For future research, considerations to explore lessons learned about communication processes and activities in Rwanda would inform new theorization about development communication in post-conflict settings. We need more research on women's communication activities in South

Sudan, Liberia, Arab Africa as well as the Congo among other conflict and post conflict societies.

Concluding Remarks

Overall, peace activities emerge as forums where women demonstrate critical positioning of their role as peacemakers in post-genocide Rwanda. In spite of the burden of cultural barriers that Rwandan women have had to overcome, the gains speak to their integrity as peacemakers. The decision to culturalize peace activities has resulted in mandating the rewriting of peace and development narrative to conform to local experiences of those most affected by the 1994 genocide. The unifying aspect of women's peace initiatives lies in joining the individual, collective and the structural (this refers to how culture, gender, memory, politics, communication, peace and development work together). Subsequently, it legitimizes women's actions and above all permits identification with the experience of the post-genocide life and vision for the future of Rwanda. The courage of women to spearhead peace building efforts has legitimized their role in almost every sphere of life in Rwanda.

Addendum

Information about the women's groups discussed is based on interviews of the groups' members by the lead author of this chapter. She has conducted research on women and peace communication in Rwanda since 2004.

Notes

1. Chen, "Toward Transcultural Understanding" in *Transcultural Realities*, ed. Milhouse, Asante, and Nwosu, 55–70. Thousand Oaks, CA: Sage, 2001.
2. Ki-Moon, "Women Integral for Peace." United Nations. www.Un.org /apps/news/story.asp?
3. Powley, "Strengthening Governance." Hunt Alternatives Fund. www. Huntalternatives.org/ download/.
4. Uwineza and Pearson, *Sustaining Women's Gains*. The Institute for Inclusive security, 2009.
5. Verwimp, "Death and Survival," *Population Studies* 58 (2004): 233–245.
6. Cohen et al., "Prevalence and Predictors," *Journal of Women's Health* 18 (2009): 1783–1791.

7. Betts, "Should Approaches to Post-Conflict," *The European Journal of Development Research* 17 (2005): 735–752.

8. Moemeka, "Interpersonal Communication in Communalistic" in *Communication in Personal Relationships across Cultures*, ed. Gudykunst, Ting-Toomey, and Nishida, 197–214. Thousand Oaks: Sage, 1996.

9. Mutua-Kombo, "Theorizing Post-Conflict," *Africa Media Review* 17 (2009): 9–22.

10. Gumucio–Dagron and Tufte, *Communication for Social Change Anthology*. South Orange, NJ: CSFC Consortium, 2006.

11. Moemeka, "Development Communication: Historical and Conceptual." In *Communicating for Development*, ed. Moemeka, 3–22. Albany, NY: State University of New York Press, 1994, 5.

12. Moemeka, "Interpersonal Communication," 197–214 and Gyekye, *African Cultural Values*. Philadelphia, PA: Sankofa Publishing, 1996.

13. Gyekye, *African Cultural Values*, 173.

14. Ibid, 171.

15. Ibid, 175–176.

16. Moemeka, "Interpersonal Communication," 199.

17. Ibid, 202.

18. Ibid, 203.

19. Ibid, 202.

20. Uwineza and Pearson, *Sustaining Women*, 12.

21. Scott, "The Evidence of Experience" in *Feminist Approaches to Theory*, ed. Hesse-Biber, Gilmartin, and Lydenberg, 79–99. NY: Oxford University Press, 1999.

22. Nzegwu, "Women and Political Leadership," *JENDA: A Journal of Culture and African Women Studies* 9 (2006): www. JendaJournal.com/issue9/.

23. McCrummen, "Women Run the Show." The Washington Post. www.Washingtonpost.com/wp-dyn/ content/article/2008/10/26/.

24. Izabiliza, "The Role of Women." UNESCO (2003). www. Unesco. org/new/fileadmin/Multimedia/hq/Shs/pdf/Role-Women-Rwanda.pdf.

25. Africa Rights, *Rwanda not so Innocent*. London, UK: Africa Rights, 1995.

26. Huesca, "Naming the World" in *Communication for Social Change Anthology*, ed. Gumucio-Dagron and Tufte, 528–535. South Orange, NJ: CFSC Consortium, 2006, and White, "The Concept of Participation" in *Communication for Social Change Anthology*, ed. Gumucio-Dagron and Tufte, 482—486. South Orange, NJ: CFSC Consortium, 2006.

Bibliography

Africa Rights, *Rwanda: Not So Innocent: When Women Become Killers*. London, U.K.: Africa Rights, 1995.

Betts, A. "Should Approaches to Post-Conflict Justice and Reconciliation be Determined Globally, Nationally or Locally?" *The European Journal of Development Research* 17 (2005):735–752.

Chen, G. "Toward Transcultural Understanding: A Harmony Theory of Chinese Communication." In *Transcultural Realities: Interdisciplinary Perspectives on Cross-Cultural Relations,* ed. V. Milhouse, M. Asante, and P. Nwosu, 55–70. Thousand Oaks, CA: Sage Publications, 2001.

Cohen, M., et al. "Prevalence and Predictors of Posttraumatic Stress Disorder and Depression in HIV-Infected and At-Risk Rwandan Women." *Journal of Women's Health* 18 (2009): 1783–1791.

Gumucio-Dagron, A., and Tufte, T. *Communication for Social Change.* South Orange, NJ: CFSC Consortium, 2006.

Gyekye, K. *African Cultural Values: An Introduction.* Philadelphia, PA: Sankofa Publishing Company, 1996.

Huesca, R. "Naming the World to Theorizing its Relationships: New Directions for Participatory Communication for Development." In *Communication for Social Change Anthology: Historical and Contemporary Readings.* ed. A. Gumucio-Dagron and T. Tufte, 528–535. South Orange, NJ: CFSC Consortium, 2006.

Izabiliza, J. "The Role of Women in Reconstruction: Experiences of Rwanda." UNESCO (2003). www.Unesco.org/new/fileadmin/Multimedia/Hq/shs/pdf /Role-Women-Rwanda.pdf.

Ki-Moon, B. "Women Integral for Peace." United Nations (2010). www. Un.org/apps/news/story.asp?NewsID=36526&Cr=&Cr1=

McCrummen, S. "Women Run the Show in a Recovering Rwanda." The Washington Post (2008). www.Washingtonpost.com/wp-dyn/content/article/ 2008/ 10/ 26/_AR2008102602197.html.

Moemeka, A. "Interpersonal Communication in Communalistic Societies in Africa." In (Ed.), *Communication in Personal Relationships Across Cultures,* ed. W. Gudykunst, S. Ting-Toomey, and T. Nishida, 197–214. Thousand Oaks, CA: Sage Publications, 1996.

———. Development Communication: A Historical and Conceptual Overview. In *Communicating Development: A New Pan-Disciplinary Perspective,* ed. Moemeka, A., 3–22. Albany, NY: State University of New York Press, 1994.

Mutua-Kombo, E. "Theorizing Post-Conflict Peace Communication: Can Rwandan Women's Narratives of Sexual Violence become the Point of Departure for Research?" *Africa Media Review* 17 (2009): 9–22.

Nzegwu, N. "Women and Political Leadership." *JENDA: A Journal of Culture and African Women Studies* 9(2006): www.jendajournal.com/issue 9/ nzegwu1.html.

Powley, E. "Strengthening Governance: The Role of Women in Rwanda's Transition." Washington: Hunt Alternatives Fund (2003). www.Huntalternatives.org/download/10_strengthening_governance_the_role_of women_ in_ rwanda_s_transition.pdf.

Scott, J. "The Evidence of Experience." In *Feminist Approaches to Theory and Methodology: An Interdisciplinary Reader,* ed. S. Hesse-Biber, C. Gilmartin and R. Lydenberg, 79–99. NY: Oxford University Press, 1999.

Ugboajah, F. "Cultural Factors in Communication for Third World Development: The Africa Case." *Communication Socialis, Year-book,* 6. (1987): 35–47.

United Nations Development Fund for Women. Women at Peace Table Make a Difference. UNDP (2000). www.Iknowpolitics.org/files/WomenAtPeaceTable.pdf.

Uwineza, P. and Pearson, E. *Sustaining Women's Gains in Rwanda: the Influence of Indigenous Culture and Post-Genocide Politics*. The Institute for Inclusive Security, 2009.

Verwimp, P. "Death and Survival during the 1994 Genocide in Rwanda." *Population Studies,* 58 (2004): 233–245.

White, S. "The Concept of Participation: Transforming Rhetoric to Reality." In *Communication for Social Change Anthology: Historical and Contemporary Readings*. ed. A. Gumucio-Dagron and T. Tufte, 482–486. South Orange, NJ: CFSC Consortium, 2006.

Chapter 12

Coding for Development in the Silicon Savannah: The Emerging Role of Digital Technology in Kenya

George Gathigi and Ernest Waititu

This chapter examines diffusion and adoption of new information and communication technologies (ICTs) in Kenya, in particular mobile telephony and web-based technologies. It analyzes their implication on development communication processes. We note the shift from the traditional top-down approach in information provision and sharing to more complex vertical-horizontal processes that bring together multiple players on a single digital platform that performs various functions at once. Using examples of mobile banking, health, and animal management applications, we posit that development communication processes that are inspired by digital technologies distribute power among different players. This facilitates communication and transactional activities while presenting new possibilities to populations. We note the growing power of the individual to create and manipulate digital platforms. We further appreciate the ability of ICTs to significantly impact the rural low-income population by providing users with possibilities in different frontiers of trade and finance, health and economic production through access to financial services, connecting them with health providers and experts in animal husbandry.

This paper contends that digital communication processes have resulted in invigoration of development communication in the last decade. Using the example of Kenya, a country that has been referred to as the Silicon Savannah due to the large number of digital innovations and tech startups in recent years, processes that are powered by mobile phones present the latest manifestation of application of communication in development.[1] Mobile phone technology has completely altered monetary transactions, the way patients take their medication and even the way farmers interact with their animals. Through three case studies of M-PESA (mobile banking), iCOW and Mobile Health, we illustrate how new communication technologies are driving development communication practice. These applications present the new frontier for development communication as their application is altering lives of millions of ordinary Kenyans in health, education, business, and agriculture.

In the first section of this chapter, we outline the different epochs in development communication theory and practice. The second section looks at the diffusion and adoption of new technologies in Kenya. It is followed by a discussion of ICTs and the digital divide in Africa. The fourth section analyzes the shift in access to digital technology by rural and low-income segments and its application in animal rearing through iCOW application, mobile banking through M-PESA platform and in health through M-Health. Finally, we discuss how application of digital technologies fit or depart in various conceptualization of development communication.

Over the years, communication strategies and principles have been applied in development processes in the developing world. As Singh notes, the development communication field has evolved in terms of communication technologies application, goals, perspective, and promotional agencies.[2] He identifies three periods through which the different characteristics can be noted. During the period between late 1950s, when the field emerged, and the early 1970s, there was an emphasis on the use of communication technologies that included radio, print media, and television broadcasting. This period was characterized by an instrumental outlook toward technology. The goal was to shift populations' mindset toward the "modern" and achievement-oriented thinking. These efforts were promoted by national governments and prominent individuals through central planning and trust in the state.

Between the early 1970s and 1990s, there was application of satellite technology in addition to the traditional mass media. Efforts were geared toward economic growth with an appreciation of institutional and structural constraints. National agencies, international organizations, research institutes, and multinational corporations (MNCs) were involved as promotional agencies. There was a guarded shift away from state-centered efforts to involve communities and private sector. The late 1990s signaled the application of new technologies, in particular computers and the internet, in addition to satellite and traditional mass media in development. This period is characterized by pluralistic, strategic, and participatory approaches to development. The goals encompass non-governmental organization (NGOs) and community-based organizations (CBOs) in empowering and networking and facilitating economic growth and private profit. Development has been promoted by national governments, international NGOs, MNCs and private sector. The dominant policy instruments have included privatization, donor agency funding and state and NGO strategies with an appreciation of structural constraints and micro contexts of empowerment.

In addition, we want to note the recent shift in the application of new technologies in some developing nations' contexts. There has been an increased access to and adoption of computers, the internet, mobile technology, and the rapid expansion of telecommunication infrastructure through satellite and fiber optics in places such as Kenya. This has resulted in technological convergence. In turn, this has changed the way technologies are used and realigned roles of players such as government, private sector and funding agencies. A notable development in the application of new information and communication technologies is the emergence of a thriving space for creativity and innovation. Individuals, with limited or no institutional support, are developing ideas that respond to populations' needs without necessarily undergoing the painstaking projects development and implementation cycles. Similarly, digital technologies are driving development communication processes and practices that have conversational tone, where "receivers" act as active creators and contributors in the communication process. These technologies have also redrawn the spaces and added to the faces involved in the development communication processes. Web-based platforms such as websites, blogs and social networking sites such as Facebook are important news spaces in their own right as well as reinforcing components of traditional media such as radio print and television.

Development Communication and New Technologies

The advent of the new millennium coincided with an unprecedented growth in information and communication technologies in Africa. Kenya is a prime example of these changes. A 2011 Mobile Africa report estimated that there were over 500 million mobile phone subscribers in Africa, a growth from 246 million in 2008. The four biggest mobile phone markets in Africa are Nigeria, South Africa, Kenya, and Ghana.[3] The growth in Kenya is well captured by the International Telecommunication Union (ITU) since 2000 when there were 126,404 mobile subscribers growing to 1.1 million in 2002 and 24 million by 2010.[4] The same trends can be noted in individuals using the internet. In 2000, only 0.32 percent of the Kenya population was using internet data. The number grew to 5 percent in 2005 and by 2010, approximately 30 percent were using internet.[5]

Wilkin and Mody have noted that the arrival of a new millennium signified a transformation in human communication due to the emerging digital and convergent communication technologies.[6] They argued that for development communication specialists, the transition raised more questions than it offered solutions.[7] From the global perspective, the growth of private investment in communication industries has cost the poor, resulting in widening gaps across the nations. The growth witnessed in the ICT sector in countries like Kenya has been largely spurred by pro-technological policies such as liberalization of the market as well as demand that has ensured high rates of returns to the ever-growing pool of investors.

The patterns of growth in Kenya and its implications provide a fertile ground to re-examine the concept of development communication. In their paper, Wilkin and Mody define development communication as "a process of strategic intervention toward social change initiated by institutions and communities."[8] The field, they argue, can be "conceptualized by affirming the intentional use of communication technologies and processes to advance socially beneficial goals, focusing on the use of communication for development"[9] Following Berger, Wilkin and Mody argue that development communication offers a valuable opportunity to consider how academic scholarship bridges institutional practice. This scholarship has moved beyond the cold war administrative and development assistance origins by recognizing the contributions of social theory, political-economic inquiry, and cultural studies. "Incorporating these theoretical frameworks beyond the traditional models of media effects, to consider the importance of political-economic conditions, organizations contexts, communication processes, and more in the processes of social change"[10] It would be good to add that development communication should capture the implications of social change that go beyond institutions and communities to recognize the increased role of individuals in utilization and creation of ICT-based platforms.

In examining the application of communication in development in Kenya today, we take a particular note of new dimensions in the processes and adoption of ICTs, which defy the traditional thinking around communication and development. First, there has been a radical shift in the roles of government and other institutional players such as international and nonprofit organizations. Traditionally, the government played a central role in facilitating the communication infrastructure and subsidizing the delivery of information. In the media effect-based interventions through radio and television in Kenya, the messaging was mainly done through the national broadcaster, which either provided free or subsidized airtime for communication programs. Since the liberalization of the airwaves in Kenya in the 1990s, the government has moved away from direct control of communication infrastructure to facilitating.[11] In the instances where the government has retained considerable interest, such as radio and television through the state bodies such as Kenya Broadcasting Corporation (KBC), the effectiveness of these platforms has been neutralized because of the tremendously diminished audience reach. At the same time the intense competition from private broadcasters has made it necessary for institutions such as KBC to generate revenue and as such, those willing to put their materials on the stations have to pay market rates. At this front, KBC has not competed favorably. In mobile and internet technology, the government's role has been regulatory, through provision of licenses. Mobile service provision in Kenya is through private companies (although the government owns shares in leading provider entities), and the whole expansion process has followed the business model with services being provided to those who can "afford."

The second dimension of the shift is the high involvement of low-income populations as a viable and stable consumer base. The consumer driven diffusion and adoption that is spurred by demand has resulted in a relatively sustainable model even among the low-income population, a group that is usually shunned by investors. Mobile telephony has completely changed the Kenyan business model by demonstrating that the low income segment is indeed profitable as a critical mass. While mobile providers started by targeting high- and middle-earning segments, the consequent expansion has targeted the predominant rural dwellers. This expansion has been augmented by presence of cheap mobile handsets and has necessitated adoption of low-price packaging of mobile products such as the ability to buy prepaid calling units (mobile airtime) in small denominations. As a result, this has made it possible for low-income subscribers to access services. The expansion of digital technology in the rural areas has gone hand in hand with traditional media, in particular radio and similar thinking in the banking industry with synergistic results. Third, the application of new technologies has borne collective processes, both coordinated and uncoordinated, where innovators, service seekers, consumers and service providers have all converged to respond to different needs leading to solutions that are based on digital platforms. The financial industry, service providers such as electricity distributors, mobile phone companies and technological innovators for example all collaborate to ensure that customers can pay their bills or recharge electricity units at home through mobile phones. Innovative individuals (developers) have been coming up with apps (mobile phone applications) that respond to specific needs in fields such as farming, business, and health, among others.

ICTs, Digital Divide and Potential in African Context

Discussion on ICTs in developing countries such as Kenya has predominantly revolved around digital divide and the unrealized potential. ICT discourse in the past has addressed the potential benefits of digital revolution.[12] Some of these benefits are: direct contribution to the economy, improved public sector administration including transparency in procurement process for public service, improvement in education and training, improvements in service delivery in healthcare and improved environmental stability, improvement in manufacturing, and overall poverty reduction.[13] ICTs and technology transfer can promote and "move a country ahead in economic, human capital, and social development."[14] Technology constitute the latest knowledge, skills and practices of economic development processes.[15]

Despite these potential and some realized benefits, scholars have noted the varied degrees of participation in the digital revolution and modern technology. There are also differences in keeping up with the rapid changes in ICTs. Most notable is the slow rate of transfer of ICTs into many Least Developing Countries (LDCs). However, there are many "deviant" LDCs that have rapidly adopted ICTs despite having less impressive economic indicators. The manifest gap between high and low ICT participation has been referred to as "digital divide." Global digital divide is evident between developed and transitional economies on one hand and least developed countries on the other. Indicators of digital divide include the number of: internet hosts, internet users, households that own computers and the number of cell and mobile phone users.[16]

At the global level, the limited application and diffusion of ICTs in some parts of the world is worsened by factors such as language. Eighty percent of world websites for example are in English. But the language is only understood by about 10 percent of the world population.[17] Of those who understand English, only very small section constitute the world poor. One outcome of the digital divide is that it worsens social and economic disparities and it enables the advantaged section of the population such as the educated to become richer while the non-educated become poorer. Scholars have noted the centrality of information access. Stewart argues that in a networked society, that is countries with highest indicators of ICTs, economic wellbeing is information and knowledge based.[18] Knowledge, he argues, is more powerful than natural resources, manufacturing infrastructure or "fat bank rolls." Accordingly, a high level of adoption and diffusion of ICTs can support economic and social developments and interactions in a society. In the digital era, ICTs are crucial tools for a country's immediate and long-term development. ICTs play a role in a country's improvement of social policy areas such as poverty, health care, agriculture and education.[19]

Two factors differentiate highly technological countries from those that are not. First, an economic environment that encourages investment and a second, a climate for civil liberties conducive to research and expansion of communication.[20] Due to absence of these conditions in many LDCs like Kenya, there is usually a "market efficiency gap," which is the difference between the level of service penetration that can be reached under the existing plans and conditions, and the level one would expect under optimal market conditions. In turn, this leads to "access gap" or lack of affordability of ICTs.[21] The relevance of ICTs among "world poor" has, therefore, been viewed with skepticism. In 2000, the World Economic forums noted that:

> On every relevant measure—from the size and penetration of telecommunications market or the extent of internet diffusion to level of global electronic commerce and so on—the vast majority of economic activity related to information and communication technologies is concentrated in the industrialized world. Conversely, developing countries, and especially least developed countries, account for a small fraction of the global digital economy.[22]

This early blanket techno- pessimistic view towards LDCs led to conclusion that ICTs are irrelevant to world population that live below the poverty line. A large majority of these individuals are to be found in Sub-Saharan Africa and South Asia.[23] Even with significant increase in the application and diffusion of ICTs in low income nations like Kenya, a lingering question has been "how will the poor benefit?" Viewed from digital divide perspective, the population segment that has access to technology derives benefits. Regarding the state of ICTs in LCDs, Rice concludes that:

> It is important for LDCs to eliminate or reduce several significant barriers that impede technological transfer and technology development within their borders. These barriers can be placed into at least four categories: economic and financial, organizational, institutional, and human resource-related. First, LDCs have a poor economic base and low incomes, which provide little stimulus for savings and investment. Second, there are few or no incentives to encourage business and market development for technologies. Third, business and market structures in LDCs must advance from a monopoly or oligarchy to one of competition and fair pricing.[24]

These conclusions are familiar and point to the widely held argument that low-income populations do not have the economic potential that allows for investment and consequently sustained information transfer. While they may be true elsewhere, the Kenyan experience points to a different direction. Instead, when individuals in low income population are provided with services that they need and in realistic measure, they are capable of providing business at sustainable levels. There is no lack of incentives that encourage business and development technologies in Kenya. Different players in the technology industry have found relevant technologies that are attractive to all segments of the population. Notably, the diffusion and adoption of new technologies in Kenya has occurred in what is largely an oligarchic market. According to Communications Commission of Kenya, Safaricom controls 70 percent of the mobile market and 80 percent of internet subscriptions in Kenya.[25] This domination did not help in keeping calling prices competitive in the formative years, until rival providers entered in what became price wars. The competitive pricing ensued after the leading telecom firm in India, Bharti Airtel, acquired Zain Africa in 2010 and lowered calling rates to compete with Safaricom. Today, Kenya has some of the cheapest calling rates in the world.[26]

In the following section, we discuss three examples of application of mobile technology in financial transactions, health and farming.

iCow Phone Application

The traditional mode of sharing farm knowledge among farmers involves swapping stories about their animals and crops in village collectives or by paying visits to their friends and colleagues. When farmers need expert opinion about their animals, they will call the veterinary extension officers to their farms. Animal rearing is one of the oldest farming practices in African society. With shrinking land, most farmers have adopted modern farming that allows them to produce more even in a small area. Messages targeting animal keepers have been part of development communication efforts using platforms such as the radio to address wide array of issues including new methods in production, feeding and nutrition, breeding, marketing, animal diseases among others.[27] Such broadcasts have been more effective than farmers' meetings as they allow sharing of knowledge among a wide audience. Occasionally, these programs invite leading experts and successful farmers to come share their accumulated knowledge. Sometimes, other contributors, mainly farmers, join in by placing phone calls to the studio to make comments or ask questions. The advent of cellular technology has allowed for even more interactivity through text messages and phone calls. Agricultural broadcasts have become richer, as a wider range of farmers contribute to the content. Since the early 1980s, these shows have thrived from a multiplicity of topics and angles serving both crop and animal keepers. Still, these radio shows were limited in many ways. At best, they would be available for just a few minutes a week, therefore, shutting out thousands of views and tones of need-driven queries.

Mobile phone application iCow represents a different response to farmers who keep livestock needs. This application allows farmers to log individual animals on a database to receive individualized text messages about the animal's reproductive cycle on their mobile phones. The platform also includes advice for veterinary care and feeding schedules, a database of experts, and updated market rates on cattle prices. That the app has been so well received among farmers is indicative of its utility. With iCow, small-scale farmers are invading technology spaces traditionally associated with the elites to improve the productivity of their farms. The app is described as "the world's first mobile phone cow calendar."[28] In 2010, it won first prize in *Apps4Africa: Civic Challenge*. The app has come in handy for livestock farmers with quick help to various challenges they face with their animals. Beyond the mobile platform, farmers can interact with experts on the internet by logging to www.icow.co.ke or a Facebook page for timely and valuable information about their animals.

Farmers like Hosea post their questions such as one below on the website and get an answer.

> I am observing a blinding white cloudish spot on the eyes of my dearest Cow and it has lots of tears what's this and medications? What causes this?

Jos Creemers, iCow expert advisor replied:

> Dear Hosea,
> That is pink-eye. If you start treating it in an early stage it can be cured if you start too late the cow may go blind. Treatment a line (1 cm) of Opticlox from Noorbrook in each eye for five days. Or Tetracycline eye powder (Pfizer) in each eye for 5 days. Note: do not touch the eye with the syringe or the bottle. Try to make sure it falls "as it where" into the eye. Good luck.

Another farmer Njeru Wa Wang'ombe added to Hosea's question:

> @Hosea that we had a cow with same problem and with tetracycline worked. After a few weeks of partial blindness the cow recovered fully. @Jos Creemers is right. Thanks icow for this forum.

The query above depicts the nature of the conversation, a non- hierarchical approach that facilitates deeper engagement with farming topics. This information is rich in expert knowledge and also includes other farmers' advice and affirmation. Others send questions such as one below, where a couple is seeking ways to boost their economic returns and points to farmers seeking general advice.

> Dear Icow,
> I am a dairy farmer in Kakamega town and would like your help in getting a high production cow that I can use to maximise productivity. Currently, I have 3 cows that produce only 6-10 litres a day and it is getting quite frustrating to work on them with the low yield. I seek a good breed that can produce over 30 litres a day. I am also part of a group of youth interested in keeping dairy goats for economic empowerment.

> I will be glad to get your assistance in this.
> Sincerely,
> Mr. and Mrs. George Muriuki"

The reply:

> "iCow" Dear Mr and Mrs Muriuki, Please send more information on the 3 cows that you have. Age, breed, birthing, lactation and medical histories as well as your feeding program. Based on this we will be able to advise you on whether having a 'better' breed will in fact result in more milk or whether your new animal may give the same results. It would be good if you could send a picture too.

These conversations point to the pertinent role that new communication technologies are playing in the lives of many people in areas of Kenya where well-developed infrastructure and top-notch research-based expertise may be out of reach to many. Development of high utility digital solutions like iCow allows those interested in farming to move beyond cows to encompass other facets of farmers' work. Taking advantage of social networking platforms such as Facebook and text messaging service increases the reach and spurs other areas of interest such as organic farming, fish farming, growing herbs, and vegetable farming. iCow represents one of the latest examples of the highly innovative culture picking up fast in developing countries such as Kenya. These innovations are challenging age-old communication forms by combining information and education delivery as well as placing tools in farmers' hands to enable them to carry out different functions that have a direct effect on their operation and ultimately, productivity. In essence, the use of newer communication technologies have not been the prerogative of urban-based or wealthy city dwellers but have also become useful to the lives and businesses of Kenyans who reside in the rural areas and farmlands.

Digital Technology and Health

Health has been one of the leading challenges in sub-Saharan Africa and Kenya is not an exception. Health issues such as family planning, reproductive, maternal and children were some of the earliest issues addressed in the communication based efforts in Kenya. In the past decade, the rising cases of HIV and AIDS and other big public health diseases such as tuberculosis and malaria received a lot of attention most epitomized by the launch of the launch of The Global Fund to Fight AIDS, Tuberculosis and Malaria. Beyond funding, an important element of the response towards the disease was providing people with information they needed to make decisions about their health. Media such as print (posters, comic books, and flyers), newspapers, and billboards, radio, television as well as the Internet have all been used mainly to pass important information and encourage behavior change.[29] This approach to messaging has been effective in reaching a large audience but uses a general approach to reach the target. However, the message can be targeted to a specific group, for example, young sexually active adolescents, people living with HIV and AIDS and so on, but still reach a large segment of people.

The application of digital technology in health and in particular mobile phone and web-based platforms through Mobile Health (mHealth) is a notable development. The use of mobile phones for medical purposes has gained traction the world over, particularly through the application of text messaging. Text messages have been used to link patients and care providers to provide timely information especially sending reminders and thereby impacting treatment. Text message helps to improve care. Research conducted in Coastal and Western Kenya on malaria case management practices showed that sending reminders to health workers' mobile phones helped to improve adherence to treatment guidelines for outpatient pediatric malaria in Kenya.[30] The mHealth efforts have also targeted people living with HIV and AIDs (PLWAS). Studies were conducted in three clinics in resource-limited settings in Kenya where patients received weekly text messages from clinic nurse and were required to respond within forty-eight hours. Results showed that patients who received SMS support in form of reminders had significantly improved anti-retroviral (ART) adherence rates and rates of viral suppression. By investing twenty cents of a dollar per patient each month one nurse could monitor adherence of 1000 patients each month.[31]

Medic Mobile, a nonprofit organization, is using mobile technology to create connected, coordinated health systems that help communities address health challenges. In Kenya, Medic Mobile is used by organizations like Kilifi Kids, which provides mobile phones to community health workers (CHWs) in Kilifi to enable them share information with the central hospital about their patients through text messages focusing on malnourished children. Doctors can then advice what course of action to take. Texting is also used to request delivery of needed drugs for sick children from their home base. The computers behind the system are programmed to tell staff which children are in the highest risk categories and in need of immediate care based on data submitted via text messages from the CHWs. This also allows the Ministry of Health to determine, by examining CHWs communications, which health workers need re-training and which might be able to serve as mentors to less experienced CHWs.[32]

MedAfrica is another mHealth application used in Kenya to promote access to health-related information and services. The application runs on a database that contains different categories of health detailed information on doctors for various ailments, location and contacts in addition to information about hospitals. The site also provides information on first aid, symptoms of various diseases and diet tips. In 2011, MedAfrica emerged as the top application in Pivot 25, a contest that pitted twenty-five applications. The app's mission is to increase access to health-related content and services to save lives and build a healthy population. It is a fundamental service in a country where phone directories are rarely accessible and accessing health services is still a big challenge. Going forward, the platform promises to constantly populate its services to progressively build a new health ecosystem. Jackie Cheruiyot, the team leader of Med Africa says that the app is built to be a one stop shop for health information. This is achieved by aggregating health content from numerous health information platforms.

Mobile Money M-PESA

Mobile banking is a service that allows cellular phone subscribers to access a virtual account separate from a banking system. In Kenya, the most popular electronic money platform is *M-PESA* (*M* denoting mobile; *Pesa*, Swahili for money), a service launched in 2007 by the leading mobile provider, Safaricom Ltd in 2007.[33] There are estimated 14 million M-PESA users in Kenya. In 2011 the mobile money service had more transactions than what its global rival, Western Union, moved across the world.[34] M-PESA is used for different purposes. These include remittances, savings, purchase of phone pre-paid credit, bill payment, bank account payments. Innovative mobile payments systems have been developed in developing economies including Asia, Africa, Middle East and Latin America but M-PESA represents one of the first and successful product.[35] According to Jack and Suri:

> Using M-PESA, individuals can exchange cash for e-money and (talk) with any M-PESA agent across the country, and transfer these balances via SMS to any other cell phone in the country (including to sellers of goods and services), even if the recipient is not registered with M-PESA and even if the phone operates on a competitor network. De-

> positing funds is free, there is a fixed fee of 30 Kenyan shillings (about 40 cents) per SMS transfer, and withdrawal are charged according to a step function at a cost of 1–2 percent (the price is higher if the recipient is not a registered M-PESA user). These fees are deducted from users' accounts, and shared by Safaricom on commission basis with the relevant agent. No interest is earned on account balances . . . "[36]

Like mobile technology on which M-PESA is based, the service is critical to development issues due to the extent to which it has allowed a huge section of the population that was hitherto restricted in financial operation, to access a seamless exchange platform. Providing a cheaper and seamless way to send money may appear to be a simple innovation. However, the impact of M-PESA so far is enormous. It allows many Kenyan households to manage their money by opening up connection to the financial system of a population segment that was previously excluded. This is important in developing countries such as Kenya where absence of infrastructure makes it difficult to provide for financial services through formal banking. M-PESA agents have brought the services, that banks would normally be expected to provide, close to the people. Most of the outlets are found deep in grassroots levels where banks would never reach.

Mobile money has been instrumental in boosting local economies promoting growth and development across the country. The platform puts cash into people's hand, thereby facilitating money movement by providing an infrastructure of financial exchange. Mobile money has led to reduced cash in the economy. Electronic money provides an easy audit trail and, therefore, transparency for regulators. For largely low-income rural segment, M-PESA also means security. M-PESA is a secure platform that is simple to use with a customized menu that allows sending of encrypted messages to a thoroughly audited financial accounting system. It is a platform where people can store their money securely. The convenience provided by M-PESA is unparalleled. It is a system that facilitates payments between low-scale producers and consumers through reduced transaction costs. Travel costs and time related to exchange are significantly reduced. Those who travel from home in search of work, whether in cities or other areas can easily send money back home for family to pay bills. Today, individuals can send money from any part of the world to the remotest area in Kenya because mobile money systems like M-PESA have been linked to the local banking and international remittance systems such as MoneyGram, Western Union and a host of small Kenyan targeted systems like PoaPay.[37] M-PESA service is also available through agents seven days a week while customers can transact from their phone around the clock.

Digital Technologies and Development Communication

In examining the impact of digital technology in Kenya today, we want to revisit the ideas that have shaped development communication field. Melkote argues that the theory and practice of development communication cannot be meaningfully discussed in the absence of definition of development as well as communication.[38] Conceptualization of development varies in the literature as it traverses interdisciplinary areas. Melkote identifies four main perspectives of thinking about and practicing development. The *modernization* conceptualization was based on "neoclassical economic theory and promoting and supporting capitalistic economic development. This perspective assumes that the Western model of economic growth is applicable elsewhere and the introduction of modern technologies is important in development."[39] Second are *critical* perspectives on development that challenge the economic and cultural expansionism and imperialism of modernization. They argue that political and economic restructuring should promote wider distribution of rewards and resources between and within societies. Third is the *liberation* or *monastic* perspectives that are informed by liberation theology which recognizes centrality of personal and communal liberation from oppression to self-sufficiency, an important goal of development. The fourth conceptualization is *empowerment* which emerged in the 1990s. It introduces concepts of powers and has been used in areas such as application of communication efforts in working with vulnerable groups. UNDP defines communication for empowerment as:

> [A]n approach that puts the information and communication needs and interests of disempowered and marginalized groups at the center of media support. The aim of Communication for Empowerment is to ensure that the media has the capacity and capability to generate and provide the information that marginalized groups want and need and to . . . discuss and voice their perspective on the issues that most concern them.[40]

UNDP approach can be extended to involve action oriented efforts such as those that are being facilitated by new technologies. People are not only able to shape the various discourses but are able to adapt and recreate ICT tools to serve different sections.

The application of digital technologies in Kenya especially as it relates to low-income segments fits within the different strands of development conceptualization discussed above. Their application is meeting the goals that the field has strived to achieve in the last fifty-plus years. While modernization as the dominant development paradigm hugely suffers from overt and covert biases, it remains true that modern communication technologies are an influential component in any efforts aimed at improving the living standards of populations.[41] In the era of global interconnectedness where individual mobility between spaces has dramatically increased, communication technologies are the glue that holds people together. To this end, their availability and access allows people to take advantage of the different opportunities that communication technologies present. Efforts that have allowed for some bridging of the digital divide allowing people and communities living in rural areas to access technologies, regardless of their economic position have been critical. Consequently, the consumers of development communication messages have changed. More than ever before, they can talk back to the government and to the development organizations. More importantly, these individuals have formed network communities which enable them to talk and talk back to each other. They are content producers who, beyond sharing simple experiences through user-submitted photos, videos, blogs, multimedia and reviews, are creating new apps that are solving shared development issues and concerns.

Digital technologies are contributing towards self-sufficiency among low-income segments. Increased mobile penetration boosts production through increased efficiency in their work and business. There are quantifiable savings in terms of money and time. These savings can be used to meet other needs. These action-oriented possibilities go beyond knowledge and have been effective because they are tools by themselves.

The emerging application of digital technologies in healthcare is an important step in African contexts where issues continue to place considerable burden on the national budget and governments can barely meet the needs. HIV/AIDS prevention, care, and management have particularly been hit hard. In 2012, huge cuts in donor aids that have mainly supported the HIV and AIDS efforts have been reported.[42] There are huge implications arising from inadequate funds in support of the overall HIV care and other health issues. Finding ways that complement health delivery increases the health workers' efficiency allowing them for example to deal with many patients and with improved outcomes. Use of mobile technology is also cost effective. As Lester et al. (2010) report in their study, "mobile health innovations can improve HIV treatment outcomes. Patients who received SMS support were more likely to report adherence to ART and were more likely to have their viral load suppressed below detection levels that patients who received the standard care alone."[43]

Groups targeted with SMS have also reported lower mortality rates. Lester et al. also reports positive reception of mobile technology among patients. "[T]he intervention was well received by patients . . . More mobile phone usage might have been promoted overall; study nurses reported a higher number of calls from controlled individuals compared with non-study clinic attendees"[44] This indicates diffusion of mobile use behavior in health among the patients. "Additionally, several participants reported forwarding their weekly message to non-intervention participants to share support." Such diffusion among health service seekers points to a potential in mass adoption. Related findings from a study on the effect of one way communication of text message reminders sent to health workers' mobile phone can improve the quality of malaria treatment.[45] "Text messages address health workers' forgetfulness, emphasize the clinical importance of doing task described in the messages, and increased the priority of doing tasks because the text messages represent the voice of authority of health workers' employer . . ."[46] The use of mobile technologies such as text messaging can be used to improve health performance. Text messaging cannot replace traditional package of case management but they can be used to complement existing implementation. Mobile applications can be used to strengthen different forms of treatment, dispensing and counseling services for different ailments in Africa. The simplicity of mobile texting and their low cost allows for potential wide scale implementation of different health interventions. Follow-up voice calls are also relatively cost effective.

Conclusion

An examination of the application of communication tools for development in Kenya in the last ten years indicates that the practice has extended beyond traditional media platforms. The practice incorporates changing realities in the development of information and communication technologies fuelled by high rate of diffusion and adoption of mobile technologies and the World Wide Web. The developments being witnessed in Kenya illustrate the various opportunities that digital technologies provide in regions that have lagged behind in the past. The new approaches are characterized by realignment of roles of different players, a notable one being the active role of individuals. They are no longer just consumers and passive receivers of information but increasingly producing and disseminating the same. Individuals are also actively engaged in designing communicative tools that respond to different needs and inexpensively. Ability to develop responsive communicative platforms indicates improved technical capacity. Although the creative group remains small, the widespread application of technologies is slowly allowing the diffusion of knowledge and more possibilities in the future. New communication technologies are altering lives of millions of ordinary Kenyans. These applications, whether in agriculture, health, or mobile banking, are changing development communication practice today. Beyond being accessible, these technologies have empowered people to take more control of their lives. In health, money transfers and other sectors of life these technologies are powered by network communities, which have allowed greater participation of people in shaping their own lives. People in these networks are occupying spaces previously dominated by big banks and big media organizations. They are replacing the top-down mode of communications with a richer, multidimensional exchange.

The current application of new technologies in addressing population issues points to a new approach where interventions can be deliberately targeted and monitored at the level of a single individual. The traditional approach of using media such as radio are effective in transmitting information and education to mass audience while mobile technologies allow for one to one interaction between different players and personalized platforms. New technologies cannot replace the traditional media and, therefore, a range of media forms become part of communication for development "mix" working in tandem to facilitate bottom-up articulation and sharing of information and knowledge. This has helped to increase efficiency in the use of development resources because the information is not only accessible but also produced at multiple points.

The application of new digital technologies presents development communication practitioners with a challenge to think about the best way to utilize limited funds. As noted above, media liberalization has resulted in rising costs of programming on the traditional platforms such as radio. This calls for practitioners to be more careful in choosing communication sites more carefully with needs in mind. Digital technologies today provide an opportunity of global access. International collaborations are possible through digital connections. With numerous efforts in areas of interests across Africa, Latin America and Asia going on concurrently, finding ways to share these innovations would be an important step. New technologies will continue to open up the development communication field to a wide range of communicators and audiences. Going forward, success in development communication will be determined by how well organizations or individuals can understand and exploit these new spaces and voices to produce better development communication products.

Notes

1. Perry, "Silicon Savanna." *Time* (2011, June 30). www. time.com/ time/.
2. Singh, "Communication Technology and Development." In *International and Development Communication: A 21st Century Perspective,* ed. Mody, 189– 198. Thousand Oaks, CA: Sage, 2001.
3. Rao, *Mobile Africa Report*, 2011.
4. International Telecommunications Union (ITU), *World Telecommunication/ICT Development Report.* Geneva: ITU, 2010.
5. Ibid.
6. Wilkins and Mody, "Reshaping Development Communication." *Communication Theory* 11 (2001): 385–396.
7. Ibid, 385.
8. Ibid.
9. Ibid.
10. Ibid, 386.
11. Gathigi, "Radio Listening Habits." PhD diss., Ohio University, 2009.
12. United Nations, *Report of the Meeting*. NY: Author, 2000; Commission of European Communities, *Information and Communication Technologies.* Brussels: CEC, 2001; and Rice, "Technologies and Global Digital Divide." *Comparative Technology Transfer and Society* 1 (2003): 72–88.
13. Ajuwon and Rhinet, "Level of Internet Access." *Health Information and Libraries Journal* 25 (2007): 175–185; Navas-Sabater, Dymond, Junutumen, *Telecommunications and Information Services*. Washington, DC: The World Bank, 2002; Graham, "Bridging Urban Digital Divides," *Urban Studies* 39 (2002): 33–56; and Rice, "Technologies and Global Digital Divide."
14. Rice, "Technologies and Global Digital Divide," 74.
15. Eggleston, Jensen, and Zeckhauser, "Information and Communication." In *The Global Information Technology Report*, ed. Kirkman, Cornelius, Sachs, and Schwab, 62–71. NY: Oxford University Press and Center for International Development at Harvard University, 2002.
16. Campbell, "Can Digital Divide be Contained?" *International Labor Review* 140 (2001): 19–41, and Rice, "Technologies and Global Digital Divide."
17. Organization for Economic Cooperation and Development (OECD), *Understanding the Digital Divide*. Paris: Author, 2001.
18. Stewart, *Intellectual Capital*. London: Nicolas Brealey.
19. United Nations Development Program (UNDP), *Driving Information and Communications*. NY: Author, 2000.

20. Rodriguez and Wilson, *Are Poor Countries Losing*?. Washington, DC: World Bank Group.

21. Navas- Sabater, Dymond, and Junutumen, *Telecommunications and Information Services.*

22. World Economic Forum, *From the Global Digital Divide.* Kyushu-Okinawa: World Economic Forum, 2000.

23. Pigato, *Information and Communication Technology.* Washington, DC: The World Bank, 2001.

24. Rice, "Technologies and The Global Digital Divide," 86.

25. Communication Commission of Kenya (CCK), "Quarterly Sector Statistics Report." CCK (2012). www.cck.go.ke/resc/.

26. Nyabiage, "Game of Numbers." *East African Standard* (2009, Dec. 9). www.standardmedia.co.ke/.

27. Gathigi, "Radio Listening Habits."

28. iCow, "Homepage." iCow (n.d.). www.icow.co.ke.

29. Hausmann- Muela, Munera, and Nyamongo, *Health-Seeking Behaviour*. London: London School of Hygiene and Tropical Medicine, 2003.

30. Zurovac et al., "Effect of Mobile Phone Text Message." *Lancet* 378 (2011): 795– 803.

31. Lester et al., "Effects of Mobile Phone," *Lancet* 376 (2010): 1838–1845.

32. Kilifikids, "How Our Solution Works." Kilifikids (n.d.). www.kilifikids.org/.

33. Jack, Suri, and Townsend, "Monetary Theory and Electronic Money," *Economic Quarterly* 96 (2010): 83–122; and Ejikman, Kendall, and Mass, "Bridges to Cash," (undated): www.cgap.org/gm.

34. *Daily Nation*, "M-PESA Transactions." *Daily Nation* (2011, Oct. 20). www.nation.co.ke/.

35. Mas, "Economies of Branchless Banking." *Innovations* 4 (2009):57–76; Ivatury and Pickens, *Mobile Banking and Low Income.* Washington, D.C.: CGAP/ World Bank/United Nations, 2006; and Jack and Suri, "Risk Sharing and Transaction Costs." Unpublished paper, 2011.

36. Jack and Suri, "Risk Sharing," 4.

37. Poapay allows for electronic international remittances from USA to mobile telephones in Kenya.

38. Melkote, "Theories of Development Communication." In *International and Development Communication: A 21st Century Perspective*, ed. Mody, 129–146. Thousand Oaks, CA: Sage, 2001.

39. Ibid, 130.

40. UNDP, *Communication for Empowerment.* NY: UNDP, 2006.

41. Melkote, "Theories of Development."

42. AIDSAlliance, "Alliance Report on Global Funds." AIDSAlliance (2012, Jan. 24). www.aidsalliance.org/.

43. Lester et al. "Effects of Mobile Phone," 1842.

44. Ibid, 1843.

45. Zurovac et al., "Effect of Mobile Phone."
46. Ibid, 801.

Bibliography

AIDSAlliance. "Alliance Report on the Global Funds Warns Millions Will Go Without Funding." AIDSAlliance (2012, Jan. 24). www.aidsalliance.org/.

Ajuwon, G.A. and Rhinet, L. "The Level of Internet Access and ICT Traiing for Health Information Professionals in Sub-Saharan Africa." *Health Information and Libraries Journal* 25 (2007):175–185.

Berger, B.K. "Private Issues and Public Policy: Locating the Corporate Agenda in Agenda-Setting Theory." *Journal of Public Relations Research* 13 (2007): 91–126.

Campbell, D. "Can the Digital Divide be Contained?" *International Labour Review* 140 (2001): 119–141.

Commission of the European Communities, *Information and Communication Technologies in Development: The Role of ICTs in EC Development Policy*. Brussels: Commission of the European Communities, 2001.

Communication Commission of Kenya. "Quarterly Sector Statistics RePort." CCK (2012). www.cck.go.ke/resc/.

Daily Nation. "M–PESA Transactions Surpass Western Union Moves Across the Globe." *Daily Nation* (2011, Oct. 20): www.nation.co.ke/.

Eggleston, K., Jensen, R., and Zeckhauser, R. "Information and Communication Technologies, Markets and Economic Development." In *The Global Information Technology Report 2001–2002: Readiness for the Networked World*, ed. Kirkman, G., Cornelius, P., Sachs, J., and Schwab, K., 62–74. NY: Oxford University Press and Center for International Development at Harvard University, 2002.

Ejikman, K., and Mass, I., "Bridges to Cash: The Retail end of M-PESA." (Undated): www.cgap.org/gm/document-1.9.49720/Bridging_%20 the_cash.pdf.

Gathigi, G.W. "Radio Listening Habits Among Rural Audiences: An Ethnographic Study of Kieni West Division in Central Kenya." PhD diss., Ohio University, Athens, OH, 2009.

Graham, S. "Bridging Urban Digital Divides? Urban Polarization and Information and Communications Technologies (ICTs)." *Urban Studies* 39 (2002): 33–56.

Hausmann-Muela, S., Munera, J., and Nyamongo, I.,"Health-Seeking Behaviour and the Health System Response." (DCPP Working Paper No. 14).London: London School of Hygiene and Tropical Medicine, 2003.

iCow, "Homepage." iCow (n.d.). www.icow.co.ke.

International Telecommunications Union (ITU). World Telecommunication/ICT Development Report. Geneva: ITU, 2010.

Ivatury, G., and Pickens, M. "Mobile Banking and Low Income Customers: Evidence from South Africa." Washington: CGAP/World Bank/United Nations, 2006.

Jack, W., and Suri, T. "Risk Sharing and Transaction Costs: Evidence from Kenya's Mobile Revolution." Unpublished paper, 2011.

Jack, W., Suri, T., and Townsend, R. "Monetary Theory and Electronic Money: Lessons from the Kenyan Experience." *Economic Quarterly* 96 (2010): 83–122.

Kilifikids. "How Our Solution Works." Kilifikids (n.d.). www.kilifikids .org/.

Lester, R., Ritvo, P., Mills, E., Kariri, A., Karanja, S., Chung, M., Jack, W., Habyarimana, J., Sadatsafayi, M., Najafzadeh, M., Marra, C., Estambale, B., Ngugi, E., Blake, B., Thabane, L., Gelmon I., Kimani, J., Ackers, M., and Plummer, F. "Effects of a Mobile Phone Short Message Service on Antiretroviral Treatment Adherence in Kenya (WelTelKenya): A Randomised Trial." *Lancet* 376 (2010): 1838–1845.

Mas, I. "The Economics of Branchless Banking." *Innovations* 4(2009): 57–76.

Melkote S. "Theories of Development Communication." In *International and Development Communication: A 21st Century Perspective,* ed. B. Mody, 129–146. Thousand Oaks, CA: Sage Publications, 2001.

Navas-Sabater, J., Dymond, A., and Junutumen, N. *Telecommunications and Information Services for the Poor.* Washington, DC: The World Bank, 2002.

Nyabiage, J. "Game of Numbers." *East African Standard* (2009, Dec. 9). Http:// standardmedia.co.ke/fj.php?id=2000042695.

Organization for Economic Cooperation and Development (OECD). *Measuring The Information Economy.* Paris: Author, 2002.

———. *Understanding the Digital Divide.* Paris: Author, 2001.

Perry, A. "Silicon Savanna: Mobile Phones Transform Africa." *Time* (2011, June 30). www.time.com/time/printout/0,8816,2080702,00.html.

Pigato, M. *Information and Communication Technology, Poverty, and Development in Sub-Saharan Africa and South Asia.* (Africa Region Working Paper Series No. 20). Washington, DC: The World Bank, 2001.

Rao, M. "Mobile Africa Report 2011: Regional Hubs of Excellence and Innovation." www.mobilemonday.net/reports/MobileAfrica_2011.pdf.

Rice, M. F. "The Digital Divide and Race in the United States." *Politics, Administration and Change* 36 (2001): 20–31.

———. "Technologies and the Global Digital Divide: Technology Transfer, Development and Least Developing Countries." *Comparative Technology Transfer and Society* 1(2003):72–88.

Rodriguez, F., and Wilson, E., III. *Are Poor Countries Losing the Information Revolution?* (InfoDev Working Paper). Washington, DC: World Bank Group, 2000.

Sahay, S., and Avgerou, C. "Introducing the Special Issue on Information and Communication Technologies in Developing Countries." *The Information Society,* 18 (2002): 73–76.

Singh, J.P. "Communication Technology and Development: Instrumental,

Institutional, Participatory, and Strategic Approach." In *International and development Communication. A 21st century perspective,* ed. B. Mody, 189—198. Thousand Oaks, CA: Sage Publications, 2001.

Stewart, T. A. *Intellectual Capital: The New Wealth of Organizations.* London: Nicolas Brealey, 1997.

United Nations. *Report of the Meeting of the High-Level Panel of Experts on Information and Communication Technology.* NY: Author, 2000.

United Nations Development Program (UNDP). *Human Development Report.* NY: Oxford University Press, 1999.

———. *Driving Information and Communications Technology for Development: A UNDP Agenda for Action 2000–2001.* NY: Author, 2000.

———. *Human Development Report–2001.* NY: Oxford University Press, 2001.

———. *Communication for Empowerment: Developing Media Strategies in Support of Vulnerable Groups.* NY: Author, 2006.

United Nations Industrial Development Organization (UNIDO). *UNIDO Initiative on Technology Transfer for WSSD: Assessing Needs-Promoting Actions.* UNIDO (2002). www.unido.org/.

Wilkins, K.G. & Mody, B. "Reshaping Development Communication: Development Communication and Communication for Development." *Communication Theory* 11(2001): 385–396.

World Economic Forum. *From the Global Digital Divide* to the *Global Digital Opportunity: Proposals Submitted to the G-8 Kyushu-Okinawa Summit 2000.* Kyushu-Okinawa: World Economic Forum, 2000.

Zurovac, D., Sudo, R.K., Akwale. W.S., Ndiritu, M., Hamer, D.H., Rowe, A.K., and Snow, R.W. "The Effect of Mobile Phone Text Message Reminders on Kenyan Health Workers' Adherence to Malaria Treatment Guidelines: A Cluster Randomised Trial." *Lancet* 378(2011): 795–803.

Chapter 13

Sustainable Development and Rural Access to Telecommunications in Nigeria

Chukwudiebube Bede A. Opata*

This chapter makes a case for the importance of extension of rural access in Nigeria based on the concept of sustainable development. It simultaneously presents a business case for telecommunications services and facilities providers to extend universal access to telecommunications services to rural areas in Nigeria. Nigeria, with a population of over 140 million people is the most populous country in Africa, about one out of every six sub-Saharan African is a Nigerian. Therefore the issue of extension of access to telecommunications services for persons in Nigeria goes a long way toward, in demographic terms, addressing access to telecommunications services in Africa, a continent largely on the have-not half of the digital divide. There is also a human side to the need for access to telecommunications in Nigeria. Nigeria's relatively low GDP per capita of $1,969 indicates that there is a pervasive problem of poverty which is compounded by a lack of access to basic health care especially in isolated rural areas. If, as I will argue below, access to telecommunications services improves economic development, access to health care and aggregate consumer welfare, there is an immense need for such improvements for persons living in Nigeria.

I will highlight the urban/rural divide in terms of access to telecommunications services and put forward an argument on the grounds of development for extension of access to unserved rural areas. Development is perceived as the primary goal of extension of access programs. Drawing primarily from economic

*The author is grateful to Professor John McEldowney and Professor Cosmo Graham for helpful comments and suggestions to earlier drafts of this work. All mistakes and errors remain the responsibility of the author alone.

literature, I argue that extension of access would be beneficial on a macro-economic level for the economic growth of the country and based on rather limited existing case studies of rural dwellers in other countries, would also maximize aggregate welfare for rural dwellers in Nigeria. I will argue that beyond merely being economically beneficial, it could promote sustainable development by integrating social benefits and environmental protection to the economic benefits earlier stated. Due to the reliance of liberalized telecommunications sectors on the role of businesses as services and facilities providers for the extension of access, I argue that there is a business case on the grounds of corporate social responsibility for these businesses voluntarily extending access to telecommunications in rural areas beyond their bare minimum legal and contractual obligations.

The Urban / Rural Divide

Expansion of access to telecommunications services to rural areas of Nigeria has been consistently identified in academic literature over the past two decades as a major challenge facing the Nigerian telecommunications sector.[1] The issue remains quite germane in the Nigerian telecommunications sector today. A majority of Nigerians live in rural areas, yet coverage of telecommunications services and access to telecommunications services, facilities and applications remains lopsided in favour of urban areas and cities when juxtaposed with the rural parts of Nigeria. There is no current data on the urban rural divide. However the World Bank remarked that digital mobile licensees have been slow to expand into rural areas and that at least twenty-two states in Nigeria would require incentives to attract mobile operators.[2]

As at February 2010, based on a population estimate of 140 million Nigerians the installed capacity of the telecommunications networks stood at 148,835,470 lines while actual number of active lines was 75,934,207 lines giving a teledensity of 55.06.[3] This represents significant progress since as at 2001, the country had a teledensity of less than one percent. However, this does not mean that 55.06 percent of Nigerians have access to telephone lines as some persons may have multiple telephone lines such as a fixed line in addition to a mobile line or more than one mobile line or business line(s) in addition to a personal number. In the absence of desegregated data collection by the Nigerian Communications Commission on rural and unserved or underserved areas and groups of persons it is impossible to cite current precise figures on the disparity between urban and rural areas.

However, two factors point to a possible concentration of resources and possible duplication of lines and effort in urban areas. These include the liberalization of the sector which has unleashed competitive pressures from a variety of operators that concentrate efforts on and have reached critical mass in terms of subscriber numbers in the more lucrative urban markets.[4] This combined with the fact that significant numbers of rural areas remain unserved or underserved by telecommunications services and without coverage of telecommunication or the signal of a telecommunication network operator indicates that most of the installed capacity, less than half of which is currently utilized, is situated in the more densely populated and more profitable urban towns and cities. There is, therefore, a need to extend access to the rural areas. Why this is necessary is addressed below in the segment on the rationale for rural access.

Rural access is an area often overlooked by legal scholars. A significant amount of the work in this area is from economists and information and communications technologists and has centred on universal service provision. This proved contemporary during the debates over competition versus universal service provision at the onset of liberalization programs in developed countries of Europe and America. Currently, however, the topical areas have been funding and the scope of universal service, specifically whether this includes access to contents and to new services such as broadband.[5] Concerns over the best possible mechanisms for funding universal service have also been reflected in the literature. In this regard, Africa and parts of the developing world appear overlooked in the literature which focus solely on the developed world or refer to the developing world without focusing on Africa.[6] The more immediate concern in Africa has been universal access; shared, public, community or group access to telecommunications services as opposed to universal service which emphasises individual or household access to telecommunication services. This chapter aims to make a modest contribution to the literature in this regard.

The Economic Development Argument for Rural Access in Nigeria

An important rationale used to justify programs for expansion of access or universal service programs is the infrastructure building and economic development argument. This perceives extensions of telecommunications networks and services as part of the provision of an infrastructure critical to the economic development of the country. Extension of telecommunications to rural and unserved areas expands and expedites the build-out of telecommunications network facilities and infrastructure in rural areas. Investment in network infrastructure construction to the extent that local labor and resources are used not only stimulates the economic growth and development of the rural area involved, but also contributes to the overall economic development of the country. In this regard, investment in extension of network infrastructure wields a societal gain that is not fully captured or internalized by the firm whose network is extended. Thus, spill-overs and externalities are integral to telecommunications infrastructure's contribution to economic development.

This view is influenced by the economic rationale argument based on the public good model which emphasises the total societal benefits associated with telecommunications network extended to these rural areas.[7] These societal benefits include "environmental public health, safety and welfare enhancements, classic 'positive' network externalities and other indirect or intangible subscriber benefits."[8] The public good model argues that not all the benefits of network extension are realized by the company operating the network and some of the benefits will accrue to the society at large. The total societal benefits include, in addition to the revenues generated by the company, consumer surplus, benefits of business expansion, savings of life and limb, and strengthening of cultural ties and telecommunications as substitute to travel.[9] This model creates significant incentives for governments and the state to intervene in the creation of universal service or rural service programs to ensure the actualisation of these benefits beyond the extent that services and facilities providers acting for narrow economic growth reasons may be inclined to roll out services.

Information is essential for development. Telecommunications, by facilitating the conveyance of information to persons in rural areas often isolated and lacking adequate access to the outside world, stimulates local economic growth and development. This is typically illustrated through its service as a substitute for travel. Anecdotal evidence of rural dwellers significantly boosting their incomes, by reducing wastage in products and wasted journeys with the help of information gleaned from contacts via telephone or through access to a local information service abounds internationally.[10] Apart from this, rural dwellers could earn a living by selling airtime and calls to neighbors contacting persons outside their neighborhood. The success and ubiquity of the Nigerian "umbrella people" has been noted. Improved incomes and livelihoods of rural dwellers could help stem the rural urban migration as they would be less inclined to search for greener pastures if their incomes and livelihoods are enhanced at home.

It is important to adduce empirical evidence in support of this anecdotal evidence that extension of access stimulates economic growth. I will cite below evidence from empirical macroeconomic studies which have specifically addressed the issue of causality between teledensity penetration and economic growth, to establish that in mainstream economic literature there is extensive empirical evidence illustrating the positive relationship between increased availability of telecommunications services and economic growth.

A World Bank report on the economic and social benefits of rural communications found that although reports of a correlation between telecommunication penetration and economic development have been in existence since the 1970s, it had been an open question whether telecommunication penetration was responsible for improved economic welfare or whether the reverse was the case.[11] It, however, concluded that a review of the literature showed that telecommunications diffusion positively impacted economic development. This view was not universally accepted in the heat of the pro-competition liberalization debates of the 1990s. For instance, Mueller strenuously argued that although telecommunications diffusion correlated with enhanced economic growth it was the result of the later rather than the cause.[12] He argued that telecommunication diffusion occurred as people and societies got increasingly wealthier hence the high teledensities of the developed world. For him telecommunication usage was a consequence of economic growth not a cause of economic development.

Subsequent macroeconomic studies utilising time series data from a cross section of countries have established that telecommunication penetration results in economic growth. Roller and Waverman investigated how telecommunications infrastructure affects economic growth.[13] They utilized data from 21 OECD countries over a twenty-year period from 1970–1990 to examine the impact that the spread of modern fixed line telecommunications networks may have had. To calculate this, they estimated a model which endogenizes telecommunications investment by specifying a micro model for supply and demand for telecommunications investment, while to identify economy-wide effects the micro-model is jointly estimated with micro-production equation. They found a causal relationship between telecommunications infrastructure and aggregate output and that one percent increase in telecommunications penetration rate could lead to a growth of up to 0.045 percent annually in the GDP of the country.[14] Furthermore, up to a third of the growth of OECD countries over the period studied was attributable to telecommunications. They also tested the notion that an implication of network externalities is that the impact of telecommunications infrastructure on growth may not be linear such that the impact on growth may be larger when a significant network size is achieved. They found that a critical mass effect exists in that the growth effect was significantly higher for countries with telecommunications penetration levels approaching universal service or a teledensity of over 40 percent.[15]

Torero, Chowdhury and Bedi utilized a substantial data set covering 113 countries for the period 1980 to study whether telecommunications infrastructure has an impact on economic growth and if it does, whether this varies with infrastructure and income levels across countries and regions.[16] Roller and Waverman's four-way equation framework was adopted with the study examining in addition the time series properties of the data collected and correcting for the presence of unit roots. They found a positive causal relationship between telecommunications infrastructure and GDP with estimates suggesting that a one percent increase in telecommunications penetration rate could be expected to lead to 0.03 percent increase in GDP less than the 0.45 percent found by Roller and Waverman. Their regression model for different countries revealed a non-linear effect of telecommunications infrastructure on economic growth. This was interpreted to mean that telecommunications networks reach a critical mass or threshold level of connectivity before impact on economic output becomes discernible. The growth effects were strongest between 5 and 15 percent outside which range growth effects were limited.[17]

Sridhar and Sridhar sought to analyze the penetration of ICT on economic development of developing economies taking into account the two way causation between ICT and economic growth.[18] The study was conducted utilizing data from sixty-three developing countries from 1990–2001 estimating results for both main telephone lines and mobile phones deploying a structural model which endogenizes telecommunications investment similar to that of Roller and Waverman. On average, the total teledensity of the developing countries was much lower than that observed in OECD countries. However, they found that one percent increase in teledensity of these countries increases national output by 0.14 percent[19] while estimates for mobile phones indicate that 1percent increase in mobile phone penetration could cause output to increase by 7 percent.[20]

Utilising the same method used by Roller and Waverman and the Endogenous Technical Change approach, Waverman, Meschi and Fuss conducted an expanded study using data from ninety-two developed and developing countries from 1980–2003 to test whether the introduction and rollout of mobile phone networks added to their economic growth.[21] They found that mobile telephony has a positive and significant input on economic growth and that this may be twice as large in developing countries juxtaposed with developed countries. The growth is larger in developing countries because mobile telephony which substitutes fixed in developing countries, but complements fixed in developed countries, provide the main network in developing countries. They found that a developing country which between 1996 and 2003 added an average of ten more mobile phones per 100 people would enjoy a per capita growth of 0.59 percent higher than an otherwise identical country.

These studies all examined the link between telecommunications penetration and economic growth. But development is more than just economic growth. Ul Haq's and Amartya Sen's works on human development and development as freedom respectively, highlight that there is more to development than economic growth alone. Thus, focus on economic growth alone may be perceived as narrow and far from encompassing different facets of development.

Writing in 2006, Neto, a scholar with a legal background, argued that while the empirical evidence on the link between economic growth and telecommunications penetration provides some support for provision of telecommunications services the evidence did not overwhelmingly support a universal service level of telecommunications provision. He argues that an evaluation of the relationship which views development beyond the narrow bounds of economic growth and a more encompassing perception of development as freedom makes a more convincing case for the provision of universal service degree of telecommunications services. Development as freedom, which was propounded by Amartya Sen and more pragmatically articulated by the policy work of United Nations Development Program, views development from the point of view of human functionings and capabilities. Neto argues that increased provision of telecommunications services contributes to development by improving at least four basic capabilities (the abilities to live a long and health life, to be educated, to enjoy a decent standard of living and to participate in community life) in addition to increasing economic growth. He contends that focusing on functionings and capabilities, including the capability to communicate, presents a stronger development case for providing universal access to telecommunications services than focus on economic growth alone.

More limited micro economic studies provide more vivid empirical evidence in support of extension of access to telecommunications services to rural areas by showcasing its effect on the lives of rural dwellers than macro economic studies. In 2007, Abraham tested in the Indian south western state of Kerala the assumption that mobile phones as conduits of information should lessen information asymmetries in markets thus making undeveloped, rural markets more efficient. He observed that fishermen at sea using mobile phones were able to respond quickly to market demand and prevent unnecessary wastage of fish (a common occurrence before they begun using phones). The phones helped coordinate supply and demand enabling transporters and merchants to take advantage of available price information to meet demand in undersupplied markets. He found that widespread use of mobile phones helped markets become more efficient by reducing risk and uncertainty. There were gains in productivity and in the sum of producer and consumer surplus while price dispersion and price fluctuations were reduced.

Access to telecommunication services also means that rural areas could become more accessible to public programs that are designed to improve aggregate welfare of recipients such as e-health and telemedicine, e-education and e-government. Cechini and Scott claim that ICT can reduce poverty by providing poor people with access to education, health, government and financial services and also helps poor farmers and artisans by connecting them to the markets.[22] They found that reaching the poor and realizing the potential of ICT for poverty reduction in areas of opportunity empowerment and security is a difficult task; however, ICT projects implemented by grassroots-based organizations and individuals who have the requisite incentives to work with marginalized groups can achieve encouraging outcomes.

This is very much applicable to Nigeria. For instance, there exist a growing number of telemedicine providers in Nigeria and dialing 191 on the MTN network in Nigeria puts one in touch with a teledoctor who can offer help and assistance from a distance with medical problems.[23] The ability to receive medical help even from a distance would arguably be even more valuable to rural dwellers that live in isolated areas. Participation in government e-programs including the ability to obtain information on vacancies and recruitments and download application forms would present significant improvements in information flows for rural dwellers that would otherwise have to take trips to capital cities to acquire this information and possibly other trips to collect, complete and submit these forms.[24]

Sustainable Development and Access Extension

In the preceding section, I argued that infrastructure building and development, particularly economic development, are significant reasons for extension of access. This, however, raises another question. To what extent would such economic development be sustainable? In other words, could the development occasioned by access extension of telecommunication services be categorized as constituting sustainable development? To the best of my knowledge, this question has not yet been addressed in the literature on sustainable development in Africa.[25] The link between access extension and sustainable development is explored in this section. I will argue below that not only is access extension broadly facilitative of sustainable development but care should be taken to ensure that the very process of access extension to rural areas is compatible and consistent with sustainable development.

There are various perceptions of what the concept of sustainable development entails, however the *Brundtland* Report's popular and broad conception of sustainable development as "development which meets the needs of the present without compromising the ability of future generations to meet their own needs" will be adopted here. As enunciated at Rio in 1992, this consists of three pillars: the economy, society and the environment. This entails the integration of three key elements; economic development, social development and environmental protection.[26] For development to be sustainable, these three elements must be present and simultaneously integrated. With regards to the extension of rural access, its contribution toward economic development of the rural areas and the country as a whole has been identified in the preceding section and will not be repeated here. I will, however, consider briefly how extension of rural access brings about social development in a manner consistent with environmental protection.

The facilitation of social inclusion has been identified as a rationale for extension of access. This has been recognized globally. For instance, Directive 2002/22/EC of the European Parliament and Council on universal service requires regulators, undertaking an assessment of services that are part of universal service, to consider whether lack of access to the service by a minority would lead to social exclusion by the group lacking access.

Social inclusion is even more important for rural areas which may be isolated, situated in remote or inaccessible areas or lacking in traditional access infrastructure such as road and rail network links with the rest of the country. Extension of access to telecommunication services for rural dwellers could facilitate social inclusion. It would provide a means for persons residing in rural areas to keep in touch with their relations residing in towns, cities or other rural areas thus fostering familial and cultural ties. Access to telecommunications could also make life in the rural hinterlands more tolerable for workers posted to rural areas as they can keep in touch with their family members residing in other areas.

Apart from social inclusion being a good ground for extension of access, the cost to the society of exclusion of some of its members, whether rural dwellers or otherwise, is too high to be borne by any responsible, civilised society. Just as there are societal benefits to be derived from extending access to rural areas, this point argues that the flipside of this is the high cost not only to the individual but to the society as a whole, to the extent that individuals are valuable members of society, where persons lack access to telecommunications services and applications. Lack of access to telecommunication includes in the extreme, inability to contact emergency services such as the police, fire services and ambulance services in an emergency. The resulting loss of property or limb and life in itself constitute significant grounds for extending at least access to contact emergency services to members of the public.

However, beyond social inclusion, access to telecommunications and IT services promotes social development of the rural communities receiving access through its efforts to bridge the digital divide. The bridging of the digital divide has long been recognized as a measure supportive of sustainable development.[27] Bridging the digital divide between urban and rural parts of Nigeria, while beneficial on its own, has the ancillary benefit of helping to bridge the digital divide between Nigeria and other information-poor developing countries on one hand and the developed countries on the other. Through the bridging of this digital divide, rural dwellers acquire information about the world around them, their country, their state and neighboring communities all in a faster, more detailed and ultimately less expensive means than pre-existing traditional means. Information acquired here is not limited to market or price sensitive information which has empowering economic value. Through training and education to utilize "new" ICT equipment, rural dwellers acquire new and valuable skills essential for meaningful social engagement and interaction in an increasingly globalized and interconnected world. They obtain access to government programs through e-government, to school and education through e-education, to modern health care particularly with diagnostics and consultation through e-health, and access to markets to producers and consumers beyond their little village through e-commerce; in these different ways bridging the digital divide alters significantly the life of rural dwellers. It ushers them into and integrates them with life in the twenty first century.

The extension of access to rural areas is not only consistent with environmental protection but could be facilitative of environmental conservation as well. Widespread adoption of telephones, particularly mobile phones, in most parts of Africa increases the deployment of telephones as a substitute for travel. As people from rural areas become able to talk with family members, friends and business associates at a distance, this obviates the necessity of undertaking more frequent journeys to keep in touch with persons. The journeys that take place could be better coordinated to ensure that both the host and the visitor meet each other, thus, resulting in fewer wasted journeys. In view of the fact that transportation, particularly road and rail transport in Nigeria like most parts of sub-Saharan Africa, relies mainly on carbon based fossil fuels which are harmful to the environment, the reduction in journeys caused by telephones as a substitute to travel leads to fewer carbon emissions and ultimately less damage to the environment. Consequently, extension of access to telephones results in the promotion of a more environmentally friendly means of social and business contact for rural dwellers.

But telecommunications can only serve as a cleaner alternative to road transportation if it is ultimately powered by cleaner energy sources. Telecommunication networks require vast amounts of equipment and facilities and some of these such as base stations require significant amounts of energy to keep them operational all day long, all year round. Given the inadequate supply of power by the national energy supply company when juxtaposed with immense unmet demand, it becomes clear that providing adequate energy to meet the demands of communications networks is a real challenge in Nigeria.[28] Currently, base stations and community communication centres set up to provide access to telecommunications services in unserved area, could be powered by generators running on carbon based fossil fuels. Apart from the fact that carbon based fossil fuels like petroleum products result in the emission of harmful carbon dioxide to the atmosphere contributing to depletion of the ozone layer and global warming, these fossil fuels once used cannot be recycled. With rising global oil prices and fears that the world's peak supply of oil could be reached within a generation, fossil fuels are not a sustainable way to power telecommunication networks in that they may not be easily available to future generations to power these networks.

Based on the foregoing, it becomes sensible and pragmatic to resort to cleaner non-fossil based energy alternatives such as wind and solar energy. These resources are freely occurring in nature and with respect to solar energy are excessively abundant, given Nigeria's location in the equator and the tropics. A requirement that telecommunications facilities and equipment be powered by carbon neutral cleaner energy sources would be sufficiently broad to be technologically neutral in that it does not specify the particular form of energy source to be deployed. Furthermore, it ensures that investment is made in building and powering networks that could be used by future generations as the energy technology would not become obsolete and unusable in future with the depletion of its resource base.

The main difficulty with carbon neutral alternative energy sources is that the available technology for harnessing solar and wind energy are quite expensive to acquire. Notwithstanding this initially more exorbitant upfront cost for these energy forms when compared with obtaining generators running on fossil fuels, it is my humble opinion that on the long run the carbon neutral alternatives work out cheaper. They have much less operational costs when juxtaposed with generators which require frequent and regular supply of diesel or petrol commodities experiencing recent price hikes with more rises in the future as we approach and surpass the world's peak oil. Consequently, although alternative energy sources may be more expensive to procure initially, on the long run they turn out to be cheaper than petroleum based generators. In addition, the carbon neutral energy sources do not have such negative externalities such as pollutants and carbon dioxide emissions that are a part and parcel of deploying generators running on fossil fuels.

Furthermore, in view of Nigeria's position as a poor developing sub-Sahara African country, which would be investing in cleaner energy to protect the environment and combat climate change, it would be a sensible proposition for the richer more polluting countries to fund the cost of adoption of these cleaner energy forms by poorer countries like Nigeria who will feel more the effects of climate change. In this regard, the non binding commitment undertaken by developed countries in Copenhagen (Broder, 2010) to provide limited amount of money for developing countries to use in fighting the effects of climate change is a step in the right direction.[29] It would, however, be preferable to develop a mechanism for coordinating and financing the transfer of cleaner technology to poorer countries like Nigeria that suffer effects of climate change. With access to the requisite technology, persons in these countries could use telephone networks to innovate and device cheaper substitutes for deployment in their poorer parts of the world. In this regard, it is important that funding is put in place for this technology transfer and that it is done in a cost oriented manner such that no more than a reasonable return on investment is generated for the technology transfer. This would prevent wealthier countries from making donations to fund the requisite technology transfer with one hand while recouping such funds and more through abnormal profits payable to firms transferring the technology.

The foregoing has illustrated the presence of the three pillars of sustainable development in the extension of access to rural areas. The elements of economic development, social development and environmental protection not only come together when access is extended to rural areas but they come together in an integrated and holistic way. Access to telecommunications and ICT is facilitative of and contributes to the improvement of aggregate economic welfare of rural dwellers, while aiding their social inclusion and bridging the digital divide, it simultaneously helps protect the environment by providing a greener and cleaner alternative to travel, especially when the energy sources powering telecommunications networks and facilities are carbon neutral.

But beyond the linking of economic development with environmental and societal concerns, sustainable development when applied to the telecommunications sector could also more narrowly refer to the long term sustainability of the sector as a viable sector of the economy. Viewed this way, it is somewhat akin to a conceptualisation of corporate sustainability put forward by Stefano Pogutz.[30] On this narrower perception of sectoral sustainability, the inefficiencies inherent in the implementation of the supplanted Posts Telegraphs and Telephones (PTT) Monopoly in Nigeria, its resulting failure to fund and deliver widespread availability of services reveal the un-sustainability of that model in Nigeria. Conversely, the relative success of sector liberalisation in mobilising resources and stakeholders beyond the state to consistently deliver services to an ever widening circle of subscribers is indicative of the greater sustainability of the liberalized model with respect to more efficient allocation of resources.

From Sustainable Development to Corporate Social Responsibility

A pivotal group of stakeholders to the actualization of sustainable development under the liberalization model are businesses, privately owned, that provide telecommunications services and facilities to consumers. The acceptance by this group of stakeholders of the importance of access extension to sustainable development is essential to the success of access extension to unserved or underserved areas under liberalization. The OECD described businesses' contribution to sustainable development as corporate social responsibility, while the European Commission's whitepaper on Corporate Social Responsibility (CSR) similarly aligns it closely with the concept of sustainable development.[31] Given the significant role that businesses have to play in the implementation of access extension to rural areas, it becomes imperative that if the sustainable development objectives of access extension are to be realised, businesses have to adopt access extension as an important part of their corporate social responsibility. In other words, services and facilities providers have to voluntarily accept to extend access to rural areas beyond the bare requirements of law and regulatory demands if the social and environmental imperatives of sustainable development in this regard are to be actualized.

Having considered the rationale for extension of access to rural areas from the macro perspective of sustainable development in the previous section, I turn below to briefly consider it from the micro perspective of the telecommunications companies extending access using the concept of corporate social responsibility of the firm. Viewed from micro CSR perspective, social and environmental dimensions discussed in the context of sustainable development are relevant and will not be repeated here. Suffice it to say in connection with the suggestion for adoption of greener and cleaner forms of energy to power networks and facilities that adoption of such energy forms which require significantly higher outlay of investment upfront and lower operating cost when compared with less environmentally friendly alternatives would be indicative of the investing corporation's confidence in the host state and communities as stakeholders in its investment.[32]

Notwithstanding Friedman's criticism of CSR which posits that the social responsibility of business is to be profitable for its shareholders, CSR has endured and flourished in management theory literature, the business practices of some leading firms in the world, and has engaged academics and practitioners from business, law, economics and politics.[33] CSR is perceived as being more than a charitable addition to business responsibilities but as something integral to the operations of business. Thus, Carroll envisages a dynamic and interacting pyramid of responsibilities of business which encapsulates economic responsibilities, legal responsibilities, ethical responsibilities and philanthropic responsibilities.[34] CSR, therefore, requires that the firm shall first of all be profitable, obey the law, and then respond to the legitimate expectations and desires of society.

Though it encompasses more than merely turning a profit, the economic bottom line is perceived as an integral part of the triple bottom line that CSR, as business contribution to sustainable development, has to address.[35] Environmental and social dimensions notwithstanding, it is my contention that there is a business case or economic bottom line argument to be made in terms of CSR for the extension of access to unserved rural areas. Vogel's review of literature on the business case concludes that the results do not conclusively show one way or another that CSR is always either profitable or loss making for the firm, that profitability of CSR for the firm has to decided on a case by case basis.[36] Zadek posited that there are four types of the business case for CSR while a recent review of the literature also identifies four major approaches in which the business case for CSR is stated.[37] These are, however, not hard and fast classifications and arguments for a business case can cover more than one type or take multiple approaches.

My argument on the business case does not fall neatly into any one of the identified types or approaches to the business case. Neither does it state that CSR in all its forms is always profitable for services and facility providers in Nigeria. My view is more precise than that. Its basic gist is that the extension of services to unserved and underserved areas as a voluntary social responsibility of services and facilities providers, going beyond their statutory, regulatory and contractual obligations could be quite profitable for these providers in Nigeria. This argument is necessary because all things being equal a rational operator of a telecommunications network will seek to add more subscribers to the network and the main reason put forward against extending networks to rural areas is that such extension is not viable. The view presented below seeks to challenge this reason.

My argument simply put is that for a variety of reasons, extension of access to rural areas could be profitable long term and is not necessarily a loss-making venture for telecommunications network services and facility providers in Nigeria. First, although rural dwellers may not generate enough outbound minutes of voice calls to be sufficiently considered profitable, they conversely tend to receive a higher volume of inbound calls terminated in their area as more affluent or technology savvy relations and business contacts call to check up on their rural relations or contacts. Therefore, although revenues from call origination by rural dwellers may be lower than elsewhere, this should not be determinative of the profitability of extending access to rural areas as significant revenues could be recouped from termination of calls originated elsewhere under the applicable call termination interconnection rate in force.

Secondly, telecommunications service providers, who commence service in areas initially deemed not profitable, could discover that the area becomes more profitable as existing customers improve their economic welfare and are able to spend more of their disposable income on telecommunications services. Thirdly, by initiating services in these areas first, the service providers build up brand recognition and loyalty in the area such that existing customers are likely to stay with the service provider when competitors enter the market and new subscribers are likely to join the known brand in the area. Fourthly, the rural service providers would have their image, logo, and brand colors displayed in their call centres, phone booths, recharge cards, and selling point; thereby getting significant free advertisement that they would otherwise have paid to display.

Although each of these points alone may not make extension of service to rural areas profitable, taken together they present a decent business case for the CSR of services and facility providers extending access to rural areas.[38] Indeed, the Commonwealth Telecommunications Organization observed that in some countries, service provision in rural areas has actually proved more profitable than operators initially thought.[39]

Based on these arguments it becomes clear that extension of access as a CSR for services and facilities providers is not presented here as a charitable addition to these providers original responsibilities but as an integral and I dare say, strategic way for providers conducting their businesses to provide a more socially and environmentally beneficial service while staying profitable. This requires bold commitments (sometimes long term) to the country Nigeria and the different host communities as co-stakeholders of these businesses in achieving sustainable development.

The argument that telecommunications services and facility operators should extend access to rural areas as part of their CSR has received at least one practical application. Although a number of telecommunications services and facility providers in Nigeria have plans for extension of access to rural areas, one company, MTN Nigeria Limited has linked this idea with CSR. MTN Nigeria Limited through its non-profit subsidiary, the MTN Foundation, has initiated a number of programs as a responsible corporate citizen; however, the relevant program for my case in point is its Rural Telephone Project. This project is provided in partnership with the Growing Business Foundation and the International Finance Corporation. It is aimed at empowering rural women entrepreneurs referred to as "phone ladies" to increase their incomes and improve quality of their lives. It helps the poor start earning a livelihood and with increased revenues expanding their business to boost household income. The project involves giving each "Phone lady" a mobile phone, MTN SIM, recharge cards, an external antenna if required, a solar panel phone charger and MTN promotional materials such as umbrella, t-shirts, face cap, money pouch and banner. The cost of this outlay will be re-paid over a six-month period. The credit is administered and monitored by micro-finance organizations which are community-based organizations, to ensure repayment of the loans. The project has impacted over 1,500 rural women in 200 communities within 30 states of Nigeria.[40] The project has been well received by the mobile telecommunications community.[41] An additional 3,000 people in 30 states of Nigeria were empowered in phase 4 of the project.[42]

This project implemented by a separate subsidiary of the company is a step in the right direction in so far as it was conducted as part of the company's CSR. It appears to be benevolent but not as far reaching as the integral business case for extending access to unserved areas put forward above. This notwithstanding, MTN at least, has a recent integral business plan to extend access to rural areas. The company estimates that there are 35 to 40 million unserved Nigerians living in rural areas. Their goal is to cover every village in Nigeria, however, their current plan aims to extend access to 850 villages and communities in Nigeria that have hitherto not received telecommunications services. The deal which will cost 40 million US dollars will require Huawei Technologies, deploying on MTN's behalf, base stations with the latest technologies which consume low energy and are environmentally friendly.[43] Although CSR was not cited as the basis of this business plan, this sort of bold business led decision to extend telecommunications access to rural areas by covering every single village with their telephone signal, utilizing environmentally friendly low energy consumption technology, is at the heart of it, in keeping with my business case argument of extending access to rural areas on the basis of CSR as businesses contribution to sustainable development in a manner that is integral to the operations of the corporation.

Conclusion

The extension of access to rural areas is greatly facilitative of economic development of the country. This is supported by macro-economic studies which show that improvements in the access to telecommunications services and facilities in a country aid economic growth as measured by the GDP of the country in question. It is also supported by microeconomic studies which provide a vivid portrayal of the beneficial impact of maximising aggregate welfare that access to telecommunications services have on the lives of rural dweller. Beyond economic growth access extension to rural areas, it is facilitative of sustainable development as it combines economic development with social inclusion and environmental protection.

Furthermore, the provision of access to telecommunications services to rural areas beyond their bare contractual obligations could be perceived by telecommunications operators as part of their corporate social responsibility, that is as part of their contribution as business to sustainable development. The argument for this is not based on an appeal to the philanthropic or altruistic nature of the companies; instead there is a solid business case for such access extension. This includes taking into account revenue derivable from call termination rates to rural subscribers, improvements to brand recognition, free advertising, and cultivation of a brand loyalty as uneconomic users become wealthier and remain as profitable customers.

Notes

1. Ajayi, Salawu, and Raji, "Nigeria: After a Century." In *Telecommunications in Africa*, ed. E. Noam, 163–177. NY: Oxford University Press, 1999; Akwule, "Telecommunications in Nigeria." *Telecommunications Policy* 15 (1991): 241–247; Onwumechili, "Dream or Reality." *Telecommunications Policy* 25 (2001): 219–231; and Onwumechili, "Reaching Critical Mass." *Africa Media Review* 13 (2005): 23–40.
2. World Bank, "Nigeria: Expanding Access." World Bank (2005). Http://vle.World - bank.org/bnpp/files/.
3. Nigerian Communications Commission (NCC), "Subscriber Data." NCC. Http:// .www.ncc.gov.ng/.
4. Onwumechili, "Reaching Critical Mass."
5. Hargroves, "Adjudication of Universal Funding." *Denver University Law Review* 79 (2002): 491–511; Nenova, "New Concept of Universal Service." *I/S: A Journal of Law and policy for the Information Society* 3 (2007): 117; and Wallsten, "Reverse Auctions." *Federal Communications Law Journal* 61 (2009): 373–394.
6. Ellig and Rotondi, "Outcomes and Alternatives." *Texas Review of Law and Politics* 12 (2007): 1–91; Manner, "Achieving the Goal." *CommLaw Conspectus* 3 (2004): 85; Sauter, "Services of General." *European Law Review* 33 (2008): 167–193; Spulber and Yoo, "Towards a Unified Theory." *Communications Law Journal* 61 (2008): 43–117; and Worthy, "Racial Minorities." *Hastings Communications and Entertainment Law Journal* 26 (2003): 1.
7. Organization for Economic Co-operation and Development (OECD). *Universal Service Obligations*. Paris: OECD, 1995. The public good model is contrasted to the private good model which does not emphasise network externalities but instead emphasises benefits (privately and internally) directly enjoyed by the subscriber to the telecommunications service. The private good model implies the definition of basic services at existing lowest common denominator in terms of scope and definition and any infrastructural improvements to the network beyond this is treated as non-basic to be priced at a level sufficient to recover cost of upgrading the network.
8. Ibid, 31.
9. Ibid.

10. The Economist, "Mobile Marvels." (2009).

11. Hudson, "Economic and Social Benefits." A Report to the World Bank, 1995.

12. Mueller, "Universal Service Policies." *Government Information Quarterly* 16 (1999): 353–358.

13. Roller and Waverman, "Telecommunications Infrastructure." *American Economic Review* 91 (2001): 909–923.

14. Ibid.

15. Ibid.

16. Torero, Chowdhury, and Bedi, "Telecommunications Infrastructure." In *Information and Communication Technologies for Development and Poverty Reduction*, ed. Torero and Braun, 21–63. Baltimore, MD: Johns Hopkins University Press, 2006.

17. Ibid.

18. Sridhar and Sridhar, *Telecommunications Infrastructure and Economic*. New Delhi, India: National Institute of Public Finance and Policy, 2004.

19. Ibid, 19.

20. Ibid, 21.

21. Waverman, Meschi, and Fuss, "Impact of Telecoms." (2005). www.Web.si.umich.edu/ tprc/papers/.

22. Cechini and Scott, "Can Information and Communication Technology." *Information Technology for Development* 10 (2003): 73–84.

23. Emeruwa, "Nigeria Offers Healthcare." Africanews (2009). www.Africanews.com/. There exists a Society for Telemedicine and e-health in Nigeria (SFTeHIN). Their information can be access via www.sftehin.org/.

24. Muhammad, "Making Councils Work." *Daily Trust*. Http:// dailytrust.com.ng/. An organisation NeGST (National e-Government Strategies) has been set up as a public-private partnership to push through e-government in the country.

25. Effeh, "Back to the Future." *McGill International Journal of Sustainable Development Law and Policy* 4 (2008): 133–161, and Sandkjaer, "Foreword: HIV/AIDS." *McGill International Journal of Sustainable Development Law and Policy* 3 (2007): 1–22. These works, representative of literature in the area, have focused thematic issues such as human rights and HIV/AIDS impact on sustainable development.

26. Segger and Khalfan, *Sustainable Development*. Oxford: Oxford University Press, 2004.

27. The Millennium Development Goals (MDGs) Goal 8; MDG Report 2009 pp.51–52; and the New Partnership for Africa's Development (NEPAD) p. 24 all recognize as sectoral priority the bridging of the digital divide by investing in information and communications technologies.

28. Nkanga and Ovia, "Between Power Supply." *ThisDay* (2007). www.thisdayonline.com/; Senge, "We Spend N700m." *Vanguard* (2007). www.vanguardngr.com/; and Ujah and Biniyat, "Power Failure." *Vanguard* (2007). www.vanguardngr.com/.

29. Broder, "Climate Goal is Supported." *The New York Times* (2010). www.nytimes.com/. By March 9, 2010, 107 countries including China and India signed a US-backed Copenhagen Accord, though nonbinding, that called for $100 billion spending annually to help emerging countries adapt to climate change and develop low carbon energy systems.

30. Pogutz, "Sustainable Development, Corporate Sustainability." In Corporate Accountability and Sustainable development, ed. Utting and Clapp, 34–60. Oxford: Oxford University Press, 2008.

31. OECD, *Corporate Social Responsibility*. Paris, OECD: 2001, and Commission of the European Communities, *Communication from Commission Concerning Corporate Responsibility*. Brussels: Commission of the European Communities, 2002.

32. Freeman, *Strategic Management*. Cambridge: Cambridge University Press, 2010.In this work, Freeman propounds a view of the stakeholder theory of CSR.

33. Friedman, "Social responsibility of Business." *New York Times* (1970): 32–33.

34. Carroll, "Pyramid of Corporate Social responsibility." *Business Horizons* 34 (1991): 39–48.

35. The triple bottom line here consists of the economic bottom line, the social bottom line, and the environmental bottom line.

36. Vogel, "Is There a Market Virtue?" *California Management Review* 47 (2005): 19–45.

37. Kurucz, Colbert, and Wheeler, "Business Case." In *The Oxford Handbook of Corporate Social Responsibility*, ed. Crane, McWiliams, Matten, Moon, and Siegel, 83–112. Oxford: Oxford University Press, 2008, and Zadek, *The Civil Corporation*. London: Earthscan Publications, 2003.

38. It is noteworthy that OFCOM in the UK has decided that BT should not receive any subsidy for its universal service provision in the UK for reasons not dissimilar to those given here.

39. Commonwealth Telecommunications Organisation (CTO), *Commonwealth African Rural Connectivity Report*. London: CTO, 2008.

40. MTN Nigeria, "Economic Empowerment." MTN (n.d.): www.mtnonline.com.

41. Ajakaye, "MTN's CSR Initiative." *ThisDay* (2006). www.thisdayonline.com/.

42. "MTN Foundation Empowers." *Punch* (2010). www.punchng.com/.

43. Osuagwu, "Rural Telephony." *Vanguard* (2010). www.vanguardngr.com/, and Uzor, "MTN Invests N6bn." *Business Day* (2010). www.businessdayonline.com/.

Bibliography

Abraham, R. "Mobile Phones and Economic Development: Evidence from the Fishing Industry in India." *Information Technology and International Development* 4 (2007): 5–17.

Ajakaye, T. "MTN's CSR Initiative Wins GSM Association Award." *This Day* (2006, Feb. 15). www.thisdayonline.com/.

Ajayi, G., Salawu, R., and Raji, T. "Nigeria: After a Century of Telecommunications, What Next?" In *Telecommunications in Africa*, ed. E. Noam, 163–177. NY: Oxford University Press, 1999.

Akwule, R.U. "Telecommunications in Nigeria." *Telecommunications Policy*, 15 (1991): 241–247.

Broder, J. M. "Climate Goal is Supported by China and India." *The New York Times*. (2010, Mar. 9). Http://www.nytimes.com/.

Carroll, A. B. "The Pyramid of Corporate Social Responsibility: Toward the Moral Management of Organizational Stakeholders." *Business Horizons*, 34 (1991): 39–48.

Cechini, S. and Scott, C. "Can Information and Communication Technology Applications Lead to Poverty Reduction? Lessons from Rural India." *Information Technology for Development* 10 (2003): 73–84.

Commission of the European Communities, *Communication from the Commission Concerning Corporate Social Responsibility: A Business Contribution to Sustainable Development*. Brussels: Commission of the European Communities, 2002.

Commonwealth Telecommunications Organisation. *The Commonwealth African Rural Connectivi ty Report*. London: Author, 2008.

Cronin, F.J., Parker, E.B., Colleran, E.K. and Gold, M.A. "Telecommunications Infrastructure and Economic Growth: An Analysis of Causality." *Telecommunications Policy* 15 (1991): 529–535.

Effeh, U. E. "Back to the Future: The UNDRD as a Blueprint for the Realisation of Human Rights and Sustainable Development in Sub-Saharan Africa." *McGill International Journal of Sustainable Development Law and Policy* 4(2008), 133–161.

Ellig, J. and Rotondi, J. "Outcomes and Alternatives for Universal Telecommunications Services: A Case Study of Texas." *Texas Review of Law and Politics*, 12 (2007): 1–91.

Emeruwa, V. "Nigeria Offers Healthcare via Phone." *Africa News* (2009, Oct. 8). www.africanews.com/.

Freeman, R. E. *Strategic Management: A Stakeholder Approach.Cambridge: Cambridge University Press, 2010.*

Friedman, M. "The Social Responsibility of Business is to Increase its Profits." *New York Times* (1970, Sept. 13): 32–33.

Hardy, A. "The Role of Telephone in Economic Development." *Telecommunications Policy* 4 (1980): 278–286.

Hargroves, J. "Adjudication of Universal Funding in the Telecommunications Sector." *Denver University Law Review* 79(2002): 491–511.

Hudson, H. E. (1995). "Economic and Social Benefits of Rural Telecommunications: A Report to the World Bank." World Bank (1995). Http:// usf.usfca.edu /fac-staff/ hudson/papers/ Benefits%20of%20Rural%20Communication.pdf.

Kurucz, E., Colbert, B., and Wheeler, D. "The Business Case for Corporate Social Responsibility." In *The Oxford Handbook of Corporate Social Responsibility*, ed. A. Crane, A.McWilliams, D. Matten, J. Moon and D. Siegel, 83–112. Oxford: Oxford University Press, 2008.

Manner, J.A. "Achieving the Goal of Universal Access to Telecommunications Services Globally." *CommLaw Conspectus* 3 (2004): 85.

"Mobile Marvels." *The Economist* (2009, Sept. 26).

The Millennium Development Goals (MDGs) Goal 8; MDG Report 2009. http://mdgs.un.org/unsd/mdg/Resources/Static/Data/2011%20Stat%20Annex.pdf.

"MTN Foundation Empowers 3000 Entrepreneurs." *Punch* (2010, Aug. 24). www.punchng.com/.

MTN, Economic Empowerment Portfolio. MTN Nigeria (n.d.). www.mtnonline.com/.

Mueller, M. "Universal Service Policies as Wealth Redistribution." *Government Information Quarterly* 16(1999): 353—358

Muhammad, H. "Making Councils Work Through eLGA." *Daily Trust* (2009, Jan. 5). www.dailytrust.com.ng/.

Nenova, M.B. "The New Concept of Universal Service in a Digital Networked Communications Environment." *I/S: A Journal of Law and Policy for the Information Society* 3 (2007): 117.

Neto, C. M. "Development Theory and the Foundations of Universal Access Policies." *I/S: A Journal of Law and Policy for the Information Society*, 2(2006): 365.

Nigerian Communications Commission. "Subscriber Data at a Glance (Year 2008-October 2009)." NCC (2010). www.ncc.gov.ng/subscriberdata.htm#summary.

Nkanga, E. and Ovia George, F. "Between Power Supply and Telecoms in Nigeria." *This Day* (2007, Aug. 30). www.thisdayonline.com/.

Onwumechili, C. "Dream or Reality: Providing Universal Access to Basic Telecommunications in Nigeria." *Telecommunications Policy* 25(2001): 219–231.

———. "Reaching Critical Mass in Nigeria's Telephone Industry." *Africa Media Review* 13 (2005), 23–40.

Organisation for Economic Co-operation and Development. *Universal Service Obligations in a Competitive Environment.* Paris: Author, 1995.

———. *Corporate Social Responsibility: Partners for Progress*. Paris: Author, 2001.

Osuagwu, P. "Rural Telephony: MTN Covers 350 Uncovered Villages with Smart Technology." *Vanguard* (2010, Oct. 26). www.vanguardngr.com/2010/10/rural-telephony-mtn-covers-350-uncovered-villages-with-smart-technology/.

Pogutz, S. "Sustainable Development, Corporate Sustainability and Corporate Social Responsibility: The Missing Link." In *Corporate Accountability and Sustainable Development*, ed. P. Utting and J. Clapp, 34–60. Oxford: Oxford University Press, 2008.

Roller, L. and Waverman, L. "Telecommunications Infrastructure and Economic Development: A Simultaneous Approach." *American Economic Review* 91 (2001):909–923.

Sandkjaer, B. "Foreword: HIV/AIDS and African Sustainable Development." *McGill International Journal of Sustainable Development Law and Policy* 3 (2007):.1–22.

Saunders, R.J. Warford, J. and Wellenius, B. *Telecommunications and Economic Development*. Baltimore, MD: Johns Hopkins University Press, 1994.

Sauter, W. "Services of General Economic Interest and Universal Service in EU Law." *European Law Review* 33 (2008): 167–193.

Segger, M.C., and Khalfan, A. *Sustainable Development Law. Principles, Practices and Prospects*. Oxford: Oxford University Press, 2004.

Senge, M. "We Spend N700m Monthly on Diesel: MTN."*Vanguard* (2007, Aug. 27). www.vanguardngr.com/.

Spulber, D. F. and Yoo, C.S. "Towards a Unified Theory of Access to Local Telephone Networks." *Federal Communications Law Journal* 61 (2008): 43–117.

Sridhar, K. and Sridhar, V. *Telecommunications Infrastructure and Economic Growth: Evidence from Developing Countries*. New Delhi, India: National Institute of Public Finance and Policy, 2004.

Torero, M., Chowdhury, S., and Bedi, A.S. "Telecommunications Infrastructure and Economic Growth: A Cross-Country Analysis." In *Information and Communication Technologies for Development and Poverty Reduction* , ed. M. Torero and J. Braun, 21–63. Baltimore, MD: Johns Hopkins University Press, 2006.

Ujah, E. and Biniyat, L. "Power Failure Behind 75 percent of Phone Network Failures: Ndukwe." *Vanguard* (2007, July 9). www.vanguardngr.com/.

Uzor, B. "MTN Invests N6bn in Rural Telephony Project." *Business Day* (2010, Oct. 22). www.businessdayonline.com/.

Vogel, D.J. "Is there a Market for Virtue? The Business Case for Corporate Social Responsibility." *California Management Review*, 47(2005):19– 45.

Wallsten, S. "Reverse Auctions and Universal Telecommunications Service: Lessons from Global Experience." *Federal Communications Law Journal* 61 (2009): 373–394.

Waverman, L. Meschi, M. and Fuss, M. "The Impact of Telecoms on Economic Growth in Developing Countries." University of Michigan (2005). Http://web.si.umich.edu /tprc/papers/ 2005/450/L%20Waverman- %20Telecoms% 20Growth% 20in%20 Dev.% 20 Countries.pdf.

World Bank. (2005, December). Nigeria: Expanding access to rural Infrastructure issues and options for rural electrification, water supply and telecommunications. ESMAP technical Paper091(2005,December).Http://vle.worldbank.org/bnpp/files/ TF027963ESMA8 142007113309091O5NigeriaRuralAccesslyerForWeb.pdf.

Worthy, P.M. "Racial Minorities and the Quest to Narrow the Digital Divide: Redefining the Concept of Universal Service." *Hastings Communications & Entertainment Law Journal* 26,(2003): 1.

Zadek, S. *The Civil Corporation.* London: Earthscan Publications ltd, 2003.

Index

About the Editors

Dr. Chuka Onwumechili is professor of communications serving as interim dean for the School of Communications at Howard University in Washington, DC. He is an accomplished scholar with several books and peer-reviewed journal articles. His work has involved development communication, particularly in the area of telecommunications. Two such books are *Reform, Organizational Players, and Technological Developments in African Telecommunications: An Update* and a co-edited publication titled *Communication and the Transformation of Society: A Developing Region's Perspective*. Among his several published works are issues dealing with universal access and building critical mass. He once served as vice president for the Digital Bridge Institute (DBI) in Nigeria where his charge included developing labor for the growing telecommunications industry in the country.

Dr. Ike Ndolo is an experienced scholar in the area of development communication and has consulted with both the British Department for International Development (DFID) and United Nations Children's Fund (UNICEF) in Nigeria. In 2009, he worked with DFID in helping to strengthen the Nigerian Bureau for Public Reform's capacity to disseminate public sector reform messages. Dr. Ndolo who currently heads the Department of Mass Communication at the University of Nigeria in Nsukka, Nigeria, is widely published. Additionally, he has also developed audio productions mostly in the area of development. Among these are his audio productions, sponsored by UNICEF in the areas of immunization, breastfeeding mothers, and compliance with prescription of health workers, among others.

About the Contributors

The 16 contributors to the book are experienced scholars and professionals in the field. Jan Servaes is UNESCO Chair in Communication for Sustainable Social Change at the University of Massachusetts at Amherst (USA) and edits *Telematics and Informatics*. Tom Jacobson is a Professor in the School of Communications and Theater at Temple University whose principal areas of research address media development, political communication, and digital journalism. Daniel O. Awodiya is a Professor of Communication at Suffolk County Community College, New York and has published several journal articles and books in the field. Annie N. Duru is an Assistant Professor of communication and her research focuses on gender communication, issues of women's rights, empowerment, and development. Matt Mogekwu teaches Journalism at Ithaca College, New York where he focuses on Media and Peace Building; International Communication; Press Freedom and Sustainable Development in Africa; and Capacity Building for Media Practitioners in Developing Countries. Tokunbo Ojo teaches at the School of Journalism and Communication, Carleton University, Ottawa, Canada. His teaching and research interests include communication for social change, global media governance, journalism studies, and political communication. Kehbuma Langmia is Associate Professor at Bowie State University in Bowie, Maryland with varied interests and extensive publication record focused on communication in Africa. Eddah Mutua—Kombo is at St. Cloud State University in Minnesota. Her work has focused on conflict and social change with special interest in post genocide Rwanda. Ayantu Tibeso is a graduate student in the Communication and Development program at Ohio University and her research interests include peace building, communication for social change, issues of diasporic identity, representation, and agency.

Some of the other contributors are not only scholars in the field but also based in Africa. For instance Ritchard Tamba M'Bayo is currently a professor at the American University of Nigeria, School of Information Technology and Communications and has written several books on media and development. While Cornelius B. Pratt is a professor in the Department of Strategic Communication at Temple University, Philadelphia, he also serves as an honorary visiting professor of mass communication at Bingham University, New Karu, Nasarawa State, Nigeria. His health work focuses on the challenges of health-care delivery in high-risk populations. Similarly, Bede Opata, taught law and researched communications regulation at the University of Warwick. He is a Barrister and Solicitor of the Supreme Court of Nigeria.

Contributing practitioners include Hilary Ozoh, a former university professor, who now practices as a professional with the United Nations Children Fund (UNICEF). George Gathigi is a scholar involved in social change projects with NGOs in Uganda, Kenya, Sudan, and Botswana. Genet Lakew is a practitioner with the AllAfrica Global Media group that hosts the popular online African news site *AllAfrica.com.* She focuses on themes close to Ethiopian and Horn of Africa communities at home and abroad, such as identity, migration, human rights and press freedom. Ernest Waititu works for Internews Network in Kenya. Internews is an international media development organization that seeks to empower local media worldwide.

www.ingramcontent.com/pod-product-compliance
Ingram Content Group UK Ltd.
Pitfield, Milton Keynes, MK11 3LW, UK
UKHW012037150926
475560UK00001B/49

9 780739 17614